Smart Sassy Animals

91 amazing true stories

Helping animals & connecting animal lovers worldwide

The publisher
Smarter than Jack Limited (a subsidiary of Avocado Press Limited)
Australia: PO Box 170, Ferntree Gully, Victoria, 3156
Canada: PO Box 819, Tottenham, Ontario, L0G 1W0
New Zealand: PO Box 27003, Wellington
www.smarterthanjack.com

The creators
SMARTER than JACK series concept and creation: Jenny Campbell
Compilation and internal layout: Lisa Richardson
Cover design: DNA Design and Lisa Richardson
Cover photograph: © Rachael Hale Photography (New Zealand) Ltd 2006. All worldwide rights reserved. Rachael Hale is a registered trademark of Rachael Hale Photography Limited.
www.rachaelhale.com
Illustrations: Amanda Dickson
Story selection: Jenny Campbell, Lisa Richardson, Anthea Kirk and Angela Robinson
Proofreading: Vicki Andrews (Animal Welfare in Print)
Administration: Anthea Kirk

The book trade distributors
Australia: Bookwise International
Canada: Publishers Group Canada
New Zealand: Addenda Publishing
United Kingdom: Airlift Book Company
United States of America: Publishers Group West

The legal details
First published 2006
ISBN 0-9582571-4-0

Contents

Responsible animal care

The stories in this book have been carefully reviewed to ensure that they do not promote the mistreatment of animals in any way.

It is important to note, however, that animal care practices can vary substantially from country to country, and often depend on factors such as climate, population density, predators, disease control, local by-laws and social norms. Animal care practices can also change considerably over time; in some instances, practices considered perfectly acceptable many years ago are now considered unacceptable.

Therefore, some of the stories in this book may involve animals in situations that are not normally condoned in your community. We strongly advise that you consult with your local animal welfare charity if you are in any way unsure about the best way to look after animals in your care.

You may also find, when reading these stories, that you can learn from other people's (often unfortunate) mistakes. We also advise that you take care to ensure your pet does not eat poisonous plants or other dangerous substances, and do not give any animal alcohol. In some rather extreme cases, you may even need to monitor what television channels your pet watches!

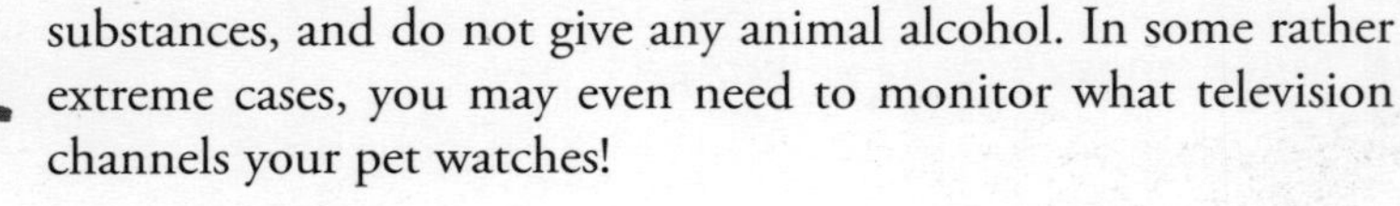

Creating your SMARTER than JACK

Smart Sassy Animals is a delightful and entertaining book that celebrates the sassy animals in our lives. You will read true stories about animals using their intelligence to have fun, outwit others, express themselves and find creative solutions.

Many talented and generous people have had a hand in the creation of this book. This includes everyone who submitted a story, and especially those who had a story selected as this provided the content for this inspiring book. The people who gave us constructive feedback on earlier books and cover design, and those who participated in our research, helped us make this book even better.

The people at the participating animal welfare charities assisted us greatly and were wonderful to work with. Profit from sales will help these animal welfare charities in their admirable quest to improve animal welfare.

Dr Hugh Wirth wrote the moving foreword, Lisa Richardson compiled the stories, did the internal layout and helped with the cover design, Rachael Hale Photography provided the beautiful cover photograph, Anthea Kirk and Angela Robinson helped with selecting the stories, Vicki Andrews did the proofreading and Amanda Dickson drew the lovely illustrations.

Thanks to bookstores for making this book widely available to our readers, and thanks to readers for purchasing this book and for enjoying it and for giving it to others as gifts.

Lastly, I cannot forget my endearing companion Ford the cat. Ford is now 12 years old and has been by my side all the way through the inspiring SMARTER than JACK journey.

We hope you enjoy *Smart Sassy Animals* – and we hope that many animals and people benefit from it.

Jenny Campbell
Creator of SMARTER than JACK

The enchanting cover photo

Masquerading behind the adoring expressions of our much loved pets are the stories and adventures, capers and escapades that have endeared them to our hearts and made them all a special part of the family.

This wonderful new edition of stories about everyday animals is brought to life with the enchanting cover photography by Rachael Hale. Rachael Hale brings you the world's most lovable animals. Friends for life. Unique characters and enchanting personalities beautifully captured to inspire and delight all ages.

With the success of this series of animal anecdotes now established in New Zealand, Australia, the United Kingdom and North America, perhaps the best story is that the sale of every book makes a generous contribution to animal welfare in that country.

Rachael Hale Photography is proud to be associated with the SMARTER than JACK book series and trusts you'll enjoy these heart-warming stories that create such cherished images of our pets, along with the delightful pictures that tell such wonderful stories themselves.

www.rachaelhale.com

The delightful illustrations

The illustrations at the bottom of this book's pages are the work of Amanda Dickson. Amanda (27) is the human retainer and devotee of her Balinese cat Miss Tilburnia Cairo Casbah (9).

When not serving Miss Tilburnia, Amanda can be found reading, watching films and, most often, drawing. Amanda is a trained cartoon animator with four years' experience in the field who has now turned her hand to book illustration.

Amanda can be contacted by email at mandsmail@gmail.com.

Connecting animal lovers worldwide

Our readers and story contributors love to share their experiences and adventures with other like-minded people. So to help them along we've added a few new features to our books.

You can now write direct to many of the contributors about your experiences with the animals in your life. Some contributors have included their contact details with their story. If an email address is given and you don't have access to the internet, just write a letter and send it via us and we'll be happy to send it on.

Throughout the book we have included other ways you can be involved with SMARTER than JACK – tell us about an amazing animal charity in your community, a smart person you know, your questions about animals' behaviour or your favourite story in this book, or send us a photo of your animal with a SMARTER than JACK book. We also welcome your letters for our 'Your say' section.

Do you like to write to friends and family by mail? In the back of this book we've included some special SMARTER than JACK story postcards. Why not keep in touch and spread the smart animal word at the same time.

Since 2002 the popular SMARTER than JACK series has helped raise over US$220,000 for animal welfare charities. It is now helping animals in Canada, the United States of America, Australia, New Zealand and the United Kingdom.

The future of the SMARTER than JACK series holds a number of exciting books – there will be ones about rescued animals and companion animals. You can subscribe to the series now too.

If you've had an amazing encounter with a smart animal we'd love to read about it. Story submission information is on page 145. You may also like to sign up to receive the Story of the Week for a bit of inspiration – visit www.smarterthanjack.com.

© Jenny Campbell (1998)

Wistful Bess, a 12-year-old Dobermann cross

Foreword

Working closely with animals and people as I do, few things surprise me. However, I am frequently amused by the free reign we allow the animals that share our lives to enjoy.

The first cat I ever owned was a male sealpoint Siamese called Lau. Like most cats, Lau could spot someone who didn't particularly like cats a mile away and my father was the perfect target. Subsequently, Lau took it upon himself to win my father's affection. He would throw himself shamelessly on my father's mercy, almost compromising his inherent cat dignity in the quest to secure a warm lap and a cuddle. His favourite trick was to crawl up inside my father's jumper until his head popped out of the neck hole. My father had little choice but to succumb and was converted from that point on.

As an animal welfare organisation, we at the RSPCA strongly encourage training and obedience to make 'good citizens' of our pets. As animal lovers, however, we are also guilty of tolerating behaviour that most would not accept from other humans – even children! A visit to any RSPCA office will likely see you encounter sleeping dogs in corridors (where they are carefully stepped around to avoid disturbing them), cats lying in desk trays and pigs, sheep and goats that have the run of the yard.

I have known people who command deference in every other area of their lives, esteemed leaders in business and the community. I have seen these same people resignedly shrug their shoulders, clearly at a total loss to explain why their pets walk all over them.

In many cases, we humans even encourage such cheekiness – who hasn't begrudgingly provided a quick scratch behind the ear before instructing a pet to get off the couch for the umpteenth time?

The big question is, of course, why? Humans are (some would say, arguably) the most intelligent creatures on earth. Why do we put up with such blatant impertinence from those that are supposed to see us as 'the boss'?

I believe the answer is that we put up with it because those humans who choose to share their lives with animals are smart. Smart enough to realise that what we get back from animals is rarely matched elsewhere. Even the most devoted admirer can barely compare with the enthusiastically affectionate greeting from a pet as you return home. Friends who will listen to our woes with that same interest and attention as the neighbourhood magpie are hard to come by. Looking into the eye of an elephant, it's difficult not to believe there is more going on inside that mind than in the minds of many who have crossed our paths.

I heartily congratulate Jenny Campbell and her team for their efforts in bringing the unique and endearing characteristics of a whole range of animals to our attention through the SMARTER than JACK series.

I would urge you to take inspiration from these heart-warming stories and look to the animals that share your life – whether they are pets, farm animals or native wildlife – to appreciate the smart, funny and cheeky things they do.

Dr Hugh J Wirth AM
President of RSPCA Australia

1

Smart animals are crafty

Nothing happened here …

My Australian cattle dog Bohdi loves to come inside the house.

One night when my partner went outside to put out the rubbish, Bohdi jumped excitedly on my lap. My partner heard him, and came in and scolded him and told him to lie down.

When my partner went back outside, Bohdi again jumped up on my lap for cuddles. This time, when he heard my partner coming back into the room, he jumped off my lap and lay on the floor and closed his eyes as if nothing had happened.

Very cheeky indeed.

Shannyn Munroe
Gold Coast, Queensland
Australia

King's trick

When I was in my teens I had a pony named King who I thought was very smart. One day the owner of the paddock I rented cut down a big tree leaving only a large stump, so I thought I would teach King to put his front legs on it and stand there as a trick. But no matter what I did, he wouldn't do it. I tried apples and carrots

King: one smart pony

and even some hay, but he still didn't even try to do the trick, so I went home very disappointed in him.

The next day when I went down to feed him, there he was with his front legs on the stump and a look in his eyes that said, *Is this what you wanted? See, I can do it on my own.*

He really was a smart pony and I will always remember the gleam in his eye – he had outsmarted me, and he did so again many times. Thanks for the laughs, King. I miss you.

Ngaire Poole
Rangiora
New Zealand

Toast of the camp

During the Vietnam War, the National Guard – a division of the US home forces – met for summer camp at Fort Hood, Texas. For two weeks, the troops lived in large tents erected in the fields among the sand and rocks and trees.

There were generators and therefore lights, and there was kerosene and therefore ovens and grills, but there were no pantries. And so the cooks had to figure out how to outwit the raccoons who came out of the trees each night to see what there was for supper.

There must be a reason why humans have been able to accomplish endeavours such as talk and wars, and raccoons have not. And one of the basic achievements of human civilisation is knowing how to keep your grain free of pests. After all, the Egyptians used cats along the Tigris and the Nile to defeat the rats – which was how humans were able to settle there and farm – and we're much smarter nowadays than those old Egyptians. So the National Guard cooks decided to use a modern practice – carpentry – to defeat a primitive enemy. They set about building a contraption to stave off the voracious racoons.

First the cooks made a framed cage screened in with chicken wire. The cage was set up on legs so it would be three feet above the ground-dwelling critters. Then they sanded the wood and lacquered it. They invited any troops who happened by to come and see their brilliant weapon against the beasts.

That's the way humans are – they see a problem and they engineer a solution. Boy, makes you proud to be one of the same species, eh? The cooks stored their bread in the cage, padlocked the gate and retired to their buddies, beer and bunks.

It took the raccoons about 15 minutes to defeat the cooks' 'pantry'. They studied it for maybe a full ten seconds. Then they knocked the whole contraption off its feet. When it fell to the ground the bread all slid to one side. The racoons had anticipated this (although apparently the cooks had not). They reached their little paws between the wire openings and squeezed and pulled the loaves through. They pulled the plastic and the bread through the fine mesh, and they chomped down every bit of the bread and left the plastic wrapper scraps strewn all about.

In the morning, the cooks set about very quickly removing the evidence of the lost battle. Then they had to think about how to adjust the breakfast menu so their defeat wouldn't be too noticeable.

Maybe spinach in place of toast?

Timothy Bowden
Felton, California
United States of America

Sneaky mouser

A couple of months ago, my friend went on holiday and asked me to look after her two pet mice. I agreed, not realising that my cat Marty was going to be a problem. You see, my cat was old and slept all day in the front yard right next to a pair of pigeons, so I figured the mice would be safe.

My cat had an excellent sense of smell and, within a day of the mice arriving, he'd found them in my bedroom. So for the next couple of days I was kept busy making sure he couldn't get near them.

One day I was getting ready to go out with some friends and, as I was checking my wallet, noticed my cat hovering round the doorway. 'Marty!' I warned him, and he turned and pretended to

Marty is cunning when it comes to mice

be sniffing my schoolbag. Once I'd turned my attention back to my wallet, he was once again staring at the mice from the doorway. 'Marty!' I said again, and he turned and ran down the stairs.

I headed out of my room, pausing for a couple of seconds to make sure the mouse cage was locked before turning off the light and shutting the door. But just as I shut the door I heard a plastic bag move in my room, so I quietly opened the door again and peeked inside to see what had made the noise.

Almost immediately, Marty's head appeared from behind my computer (which is under my desk) to see if I had left the room.

So while I thought he'd run downstairs he'd actually tricked me, and in the few seconds it took me to check that the mouse cage was locked he'd snuck past me and hid behind my computer!

In the end, my friend let me keep her mice because I fell in love with them. And boy, with my good old cat Marty around they sure are hard work!

Vanessa Stephen
Green Point, New South Wales
Australia

One determined rat

Biggie is a gentle black and white hooded rat. He is also very intelligent – and cheeky.

We kept our pet rats in a room with the door closed. We had given up on trying to contain another of our rats, the intrepid Nugget, but we didn't want them all out at once.

Our family was sitting in the living room, when down the hall came Biggie. Since no one had opened the door, even for a second, we couldn't figure out how he kept getting out.

Photo by Christopher Hamilton

SheLa with Biggie

Determined to get to the bottom of this, we lay on the bed in the rats' room, hoping to see how Biggie got out. Biggie would approach the door but, as soon as we stuck our heads out and he realised we were awake, he would just stop and stay completely still. We had to pretend we were asleep.

Biggie tried again. This time, he looked over at us. Satisfied that our eyes were closed, he pressed his body against the door until the latch popped, then – just like a person – he used his two hands to grab onto the door and pull it open. When it opened a little, he walked out.

SheLa Morrison
Gabriola Island, British Columbia
Canada

Write to me ...
SheLa Morrison
986 Lewis Close
Gabriola Island BC V0R 1X2
Canada
or email SheLa
smorrisonhamilton@yahoo.ca

A terrier uses his Aussie ingenuity

Early Easter Sunday morning, my daughter rang to say she had left her basket of Easter eggs on the ground outside by mistake (after a family Easter dinner the night before). It had a dozen medium-sized chocolate eggs in it, and she was worried that our dog Toby may have eaten them and could be very sick.

I had a look and saw that he may have eaten one egg (there was silver paper on the ground all torn up) but he seemed okay, so I told her that the kids had probably eaten some as there were only four left. I went back to bed.

Later in the day, I heard my husband yelling at the dog and thought he must have buried the bone from the roast in our garden. To my amazement, in walked my husband with five eggs in his hand, covered in dirt but not one broken.

Toby had buried one egg in each of the five plant pots hubby had planted all his bulbs in the day before, ready for spring.

I thought Toby was pretty clever and couldn't keep from laughing. Hubby was not amused. However, he did agree later that it was good thinking for a little Aussie terrier.

Mrs Laurel Johnson
The Entrance, New South Wales
Australia

You won't catch me that easily

A stray cat looking for a new place to stay couldn't have picked a better home. With five cats already, he probably thought no one would notice another mouth to feed at dinnertime. None of the other cats seemed to mind their new friend and were quite happy to share their plates.

Both our parents were adamant that five cats was already far too many and that no one was to feed Dusty (as he became known), but someone kept feeding him. Soon he knew the times for breakfast and dinner, and it didn't take long for him to wander inside and meow loudly if his dinner was late.

Although he was happy to come inside to get dinner, he still wasn't sure if he wanted to just hang out with us. So, after noticing that he had a sore ear, we decided dinnertime was our best bet for catching him for a trip to the vet. Not to be tricked so easily, he picked up a lump of cat food and took it under the porch for dinner that night.

The next night, with my sister armed with a towel to wrap him in and me as backup, he decided to forgo dinner to get away from the cat box.

As a last try, we decided on the classic trick of a cosy towel-laden cat box containing a treat of a handful of cat biscuits as a reward. Dusty was quite happy to go up to the cat box and we were excited, thinking he was going to walk into our trap. He'd obviously seen this trick done before, as – instead of walking into the box to get the cat biscuits – he picked up the towel in his mouth and pulled the whole thing out, biscuits and all, and proceeded to eat them next to the cat box.

Rachael Shepherd
Manukau
New Zealand

Third time lucky

I confess to being a mixed-breed snob. By this I mean that I have a passion for the true Heinz 57. (Heinz 57 is a slang term for mixed-breed dogs, from the H J Heinz Company's slogan '57 varieties'.) Part of this no doubt comes from the fabulous dogs who have passed through my life who came from humble beginnings.

Cooper is quite possibly one of my favourites. He was rescued when he was a puppy by ARF (Animal Rescue Foundation) Ontario with whom I worked, but ran into problems with his adoptive family. As a puppy he demonstrated tremendous mental dexterity that resulted in strained relationships. He quickly showed an aptitude for creating trouble to gain attention.

The adoption did not work out, and I was once again required to evaluate his temperament. Although I tend to reserve judgments of intellect for humans, in this case I specifically wrote, 'This dog is above average in his ability to problem-solve. The foster home and new adopter must remain on their toes.'

Sadly, Cooper was skittish when returned to the group. During his transfer to his foster home, he wrestled his 70 pound body out of the grasp of his new foster dad and darted. That night groups of volunteers hunted the river trails to no avail. He was last seen fleeing east at lightning speed.

This dog had been given a second chance, and had lost it. Accidents like this are rare, but Cooper was no ordinary dog.

The next morning while everyone planned the next phase of 'Find Cooper', the doorbell rang at his foster home. The foster mom opened the door to a dirty but otherwise healthy Cooper.

Cooper not only got a second chance, he gave himself a third. After fleeing from a home he had been to for only moments, he'd spent the night outside. In the morning he'd found his way back. Not only did he find the house, he rang the doorbell to be let in.

Cooper is doing well several years after his rehabilitation in foster care and placement into a new adoptive home.

Yvette Van Veen
London, Ontario
Canada

Write to me …
Yvette Van Veen
c/o Meeting Milo/Awesome Dogs
133 Brisbin Street
London ON N5Z 2L9
Canada
or email Yvette
info@awesomedogs.ca

A clever escape

When I was living on a dairy farm in Sopworth in south-west England, the Duke of Beaufort's hunt came through the village on their way back to the Badminton estate after a fruitless day's fox-hunting. Exhausted hounds and horses, drenched in sweat, trotted through the high street at about 5 pm on their way home.

I was standing outside our farmhouse watching them, when I noticed something moving on the tiled roof of our cowshed. Imagine my surprise when I realised that the hunted fox was sitting on the roof, watching with me. I'm sure he was smiling!

Angela Espley
Auckland
New Zealand

Write to me ... ✉
Email Angela
espley@xtra.co.nz

A treasure cave

When my American friend Mel lived in New York, she had a ferret called Zuri. My Australian friend, also named Mel, was visiting her.

Aussie Mel found that her things were going missing but thought no more about it. Then one day during the visit both Mels noticed Zuri scurrying back and forth across the lounge room. They decided to investigate, and watched as Zuri busily burrowed into Aussie Mel's suitcase, emerged with a tube of toothpaste, slowly dragged the toothpaste across the floor, then scurried behind the couch with it before re-emerging, empty-pawed.

Photo by Melissa Wimer

Sneaky Zuri the hoarder

They checked behind the couch and found a wonderful stash of goodies: a hairbrush, stockings, a notebook, socks, underwear – in fact, anything her tiny body could manage to tug from the suitcase to the wondrous treasure cave behind the couch!

Karen Fifield
Chatswood, New South Wales
Australia

Write to me ...
Email Karen
karenfifield55@hotmail.com

Does your local charity do great work in your community?

Our mission is to help animal charities in their admirable quest to improve the welfare of animals. Do you think your local animal charity does exceptionally good work? We would love to hear about it.

The best submissions received will be published, with your name, in future SMARTER than JACK books. Animal lovers the world over will then get to see how special your chosen charity is.

Submissions should be 100–200 words. For submission information please go to page 146.

2

Smart animals play tricks

Sammy made collecting the mail fun

What mail?

We had a very smart poo-terri named Sammy who had a lively sense of humour. He enjoyed playing little tricks, especially on my dad. My parents and Sammy retired to a cottage, and every day Dad and Sammy would make a trip down the lane to the mailbox.

One day Dad gave the mail to Sammy to carry home in his mouth and back they headed up the lane. When they reached the house,

Dad discovered that Sammy didn't have the mail. He feigned worried surprise and looked around, saying, 'Sammy, where's that mail?'

Sammy hesitated for a moment, and then sped back down the lane to a spot about halfway where he had left it. He came racing back to Dad with the mail in his mouth, but then continued on past him to give it to Mom inside the house, further teasing my dad.

The first time this happened, Sammy had likely put the mail down to sniff at something and had indeed forgotten it, but after seeing my Dad's startled reaction he thought it was great fun to pretend to lose the mail every day. Dad, of course, enjoyed playing along.

Joan Barclay
Markham, Ontario
Canada

The copy-bird

For the first few years after I came to Biggenden there was an old magpie who used to bring his new family round to visit me every year. I knew it was the same old bird because he had a damaged wing – the result of a brush with some boys with stones when he was younger, according to a neighbour who'd been here longer than me.

Although I don't make a habit of feeding wild animals, I always gave him a bit of bread or biscuit as a treat when he brought his family round. But one year the old fellow often looked different – the same dragging wing, but sometimes he arrived without his family and seemed glossier and fitter-looking, despite his advanced years. Then one day the mystery was solved. There outside my door were two birds with dragging wings – the old fellow with the genuinely damaged wing, and a younger impostor who'd been smart enough to trick me into giving him treats by imitating the old fellow's dragging wing.

Sadly, that was the last year I saw the old fellow (and the impostor – apparently he wasn't able to maintain the deception without the old fellow around to copy). The old fellow's first wife must have died a couple of years previously, because in the last two years he visited me he had a younger mate. So it was no surprise when he didn't turn up one year, but I have to admit that I miss him. When a bird trusts you enough to bring his new family round to visit you every year, you can't help feeling there's something rather special between the two of you.

Peter Schaper
Biggenden, Queensland
Australia

No silly kid

As children, we had a pet goat named Giddy.

For most of the day Giddy was tethered with a long rope at the back of our section. She had to be watched when she wasn't tied up, since she enjoyed chewing the clothes hanging on the washing line and would go inside the house if she got a chance. Once she jumped onto the dining table, and another time she jumped onto the coal range while the fire was burning and landed among the pots of boiling vegetables.

Shortly after arriving home from school one day, I could hear the plaintive bleats of my pet. I could see Giddy hopelessly entangled in the rope, lying on her back with her legs kicking in the air. I ran as quickly as I could to rescue my poor little friend. I carefully untangled her, stood her on her feet, and reassured her with much

cuddling and words of comfort. I then went inside for my biscuit and after-school drink.

I couldn't believe it! Within minutes, there was my goat on her back again, crying out for help. I called out to Mum to let her know how awful it was to have my pet tied up and tangled all the time.

Mum quietly told me not to go outside, but to wait and watch from behind the net curtain at the window. I waited and watched. Before long the bleating stopped. The goat looked around to see if anyone was about, then quietly stood up, turned round a few times and, when completely untangled, continued to graze happily.

Faye Ross
Ngatea, Hauraki Plains
New Zealand

A slip-up

We have a rather cheeky three-year-old curly-coated retriever named Chilliwack. Chilli is one of the family: she lives inside with me, my husband and our four young children. There are some rules, however, and one is that she isn't allowed on the beds or the couch. Sometimes Chilli gets to stay inside on her own if I'm not out for too long.

One day I came back from shopping to find a big doggy-sized dint in one of the doonas (quilts). I had my suspicions but no proof, as Chilli was on her bed when I walked in the door. A few more incidents like this happened, such as a warm spot on the couch when I came in from gardening (the kids were at school), but Chilli would always be sound asleep on her bed.

Then one Saturday Chilli slipped up. The kids and my husband were making their rowdy way out of the house for tennis lessons. She must have heard the door slam and thought I'd gone too. I walked into the lounge a few minutes after the car had gone down the drive, to find Chilli curled up on the couch. When she saw me, the look on her face said, *Uh oh, I thought you weren't here!* She quickly jumped down and got onto her bed. I'm sure she was hoping I would think it was all a trick of the light!

Janet Blaby-Dixon
Clayton, Victoria
Australia

One smart tui

We are involved with Bird Rescue, an organisation that cares for and rehabilitates sick, injured and motherless birds. One day someone brought us a tui (a native New Zealand bird) that had flown into a window.

There didn't seem to be any major problems with the bird, so we put it into an aviary and fed it daily with honey water and fruit. Each time we went near the aviary it would be sitting on the floor with its wings drooping, looking very sad, but it was eating everything we offered it.

The tui had been with us for about three months and, as we could see no improvement, we thought it would be with us for the rest of its life. Then one day Alec went in to feed it. As usual it was sitting on the floor, so he put its food in the cage and left. As he was walking back to the house, Alec remembered he hadn't changed the water so he turned back to the aviary.

The tui obviously hadn't heard Alec coming back, as it was flying all round the aviary like there was nothing wrong with it. As soon as it saw Alec, it went straight down to the floor and sat there looking miserable.

The clever bird must have known that as long as it looked sick we would carry on feeding it!

That was the end of its stay with us. We took it back to where it had come from, and it flew back up into the trees where its mate was still waiting for it.

Ailsa Reece
Kaitaia
New Zealand

Write to me ...
Email Ailsa
fun-e-farm@xtra.co.nz

Keira's avoidance strategy

My two Staffordshire bull terriers Keira and Clancy are very friendly and have been well socialised with other dogs. They enjoy meeting other dogs on our beach walks in the leash-free beach area and will happily play with them.

We did not realise Keira had a phobia about large dogs until she encountered two Great Danes on the beach. When she spied the Great Danes heading towards her at great speed, she immediately fell over and lay motionless. Clancy quickly ran to her and started sniffing her as though asking, *What are you doing?* He had spent time with Great Danes in the past and was happy to play with these gentle giants; he seemed rather confused by Keira's action.

The Great Danes ran over to Keira to check her out and she simply continued lying on the beach, absolutely motionless. Her eyes were shut tight, but every now and then she would open one eye and take a quick look to see if the two giants were still there, then immediately shut her eye again.

Clancy and Keira at the beach

The Great Danes eventually lost interest in the motionless dog and happily took off with Clancy to run along the beach. Keira then opened her eyes and, realising they were gone, got up and started to run along the beach in the opposite direction.

Suddenly, much to Keira's horror, the Great Danes turned round and began running straight towards her again. She immediately fell over and went into the 'dead dog' routine again. This time the dogs just ran straight past her, with Clancy trying hard to keep up with them. When Keira opened her eyes and realised they were gone, she jumped up and made a beeline for me, and tried to climb up me.

What a sight we must have been – two Great Danes and one Staffy happily running along the beach, and the other Staffy clinging to me and being carried!

Cathy Gillott
Worrigee, New South Wales
Australia

A toy to impress

Our 'rescue' dog Ozzie loves toys. When we first adopted him, he didn't know what toys were and didn't understand how to play. Over the two years we have looked after him, we've taught him to play and his favourite toys are fluffy squeaky ones.

We went away for a weekend, and Ozzie went to stay with my mum for his own luxury holiday of pampering and spoiling. Mum gave him a toy while he was there, a small leather football. Ordinarily, this wouldn't interest Ozzie as it was neither fluffy nor squeaky, but to be polite he played with it during the weekend he was with her.

When he came home, the football was banished to the bottom of his overflowing toy box, never to be played with again.

About three months later my mum came to visit us. Ozzie always greets guests by getting one of his toys and running over to them for some games. He usually goes for his favourite yellow duck, which he keeps near the top of his toy box.

After he'd opened his box (yes, he opens it himself and when he's finished playing he returns his toys to it and closes it!) he searched through his toys, but couldn't seem to find what he was looking for. He took out nearly all the toys – including his favourite duck – until he found the small leather football at the bottom.

He picked this up and ran over to Mum, who was chuffed to bits (very pleased) that he loved his football so much it appeared to be the only toy he ever played with!

Amanda Day
Haywards Heath, West Sussex
England

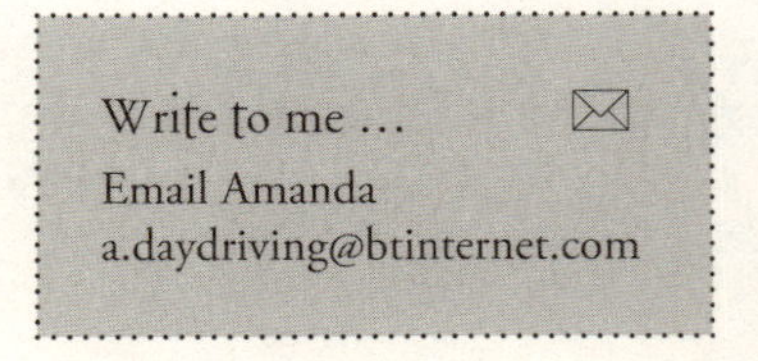

Ever seen a cat grin?

Susie was my small, pretty, tricoloured cat, with exactly half of her face coloured ginger and the other half grey. From an early age she proved to be a good hunter. When she caught her first snake (a small 30 centimetre tree snake), I made a fuss of her to make her feel important. Then, forgetting she was watching, I flung the snake a good distance away and carried her upstairs to the cupboard with the cat biscuits to give her a special treat.

Ten minutes later I went upstairs again to find her sitting erect in front of the cupboard, with the dead snake stretched out in front of her and a grin from ear to ear. 'No, I will not pay twice for one snake,' I told her. Although this happened 20 years ago now, her grin was unforgettable.

Mrs J Jurgens
Bowen, Queensland
Australia

A clever mimic

The tui, a native New Zealand bird, is a well-known mimic in the New Zealand bush. Although it has a beautiful warbling call of its own (which it often concludes with a snort or 'raspberry' sound), the tui copies other birds' calls from time to time. I was well aware of this fact, but it didn't stop me from being tricked – hook, line and sinker!

A few years ago I went for a day's walk in the Hunua Ranges, which are 30 kilometres or so south of Auckland. It was a stunning September day, and I enjoyed watching the tuis feasting on the bright yellow kowhai flowers and red flax heads. They made quite a mess of themselves doing this, their normally well-groomed blue-black plumage becoming streaked with pollen.

When I returned to the car park it was late afternoon. I unlocked the car and sat down, somewhat glad to take the weight off my feet. Hearing the beeping noise of a vehicle backing alarm gave me another reason to just sit there – I was in no hurry to leave so it was easier to let the other car finish its manoeuvring before I drove away.

It was only after I had sat there for perhaps 20 seconds that I became aware that these *beep ... beep ... beep* sounds had been going on for quite some time, and I turned round in my seat to see what the hold-up was. I was quite surprised to note there were no other cars in the parking area with people in them, let alone any that were moving.

I got out of the car quietly and saw a tui sitting in a nearby tree, 'backing' away happily. I guess it had heard plenty of real car backing alarms and had decided this was a song worthy of adding to its repertoire!

I prefer the tui's own call to the sound of a car, and I suppose it's a little sad that we have intruded on its world to that extent.

Tony Southern
Wellington
New Zealand

Cheeky boy

I opened the door to find a distraught stranger holding our dog Ratz in his arms, and my two young sons standing behind him in tears. The gentleman explained that Ratz had run out in front of his car on a busy highway while my children waited to cross the road.

Not sure what injury had been sustained, we laid Ratz on a soft bed in the lounge. After a close examination he seemed to be limping, but other than that all else seemed fine. We carried him out to the toilet for two nights, kept him warm and gave him lots of TLC.

On the third morning, while preparing breakfast, I glanced through the glass doors into the lounge room. Much to my amazement, Ratz was walking around normally – not even a limp – as if nothing had happened.

Marching into the room, I was about to tell him his game was up and 'Outside!'. To my astonishment, the moment I walked in, Ratz dropped onto his belly and began dragging himself along the floor as if he could not walk. The look on his face would have melted anyone's heart.

Who says dogs are not smart?

Marion Cook
Rosebud West, Victoria
Australia

Write to me …
Marion Cook
28/349 Eastbourne Road
Rosebud West VIC 3940
Australia

3

Smart animals have a laugh

One bird's opinion

Once I was visiting Germany and staying in a youth hostel. One evening I was alone in the dining room, just sitting and resting, when someone started to talk to me in German, which I do not understand. It was a parrot.

He kept talking to me, asking questions, for about half an hour, but I kept silent. Then he shouted, *Dumkopf!* (idiot) at me and I did not hear another word from him …

David Jenkinson
Knoxfield, Victoria
Australia

Feline mischief

I was only four when my father (Bill) brought them home: three little balls of black fur that he'd found in a sack. The three kittens had been dumped in the sewers because nobody wanted them. Bill heard them crying and climbed down into the darkness to find them.

We lived in Masterton then and Bill worked in Wellington, so he did the only thing he could think of: he put them in a cardboard box and brought them home on the train. Naturally they got out and ran amok on the carriage, but he got them home in the end.

I was delighted! Even more so when I was allowed to name them. Being my father's daughter, I was a big fan of Beatrix Potter and it didn't matter to me that these were cats, not rabbits. Nor did it bother me that there were three rather than four. I stuck to my guns and insisted they were named Flopsy Mopsy, Cottontail and Peter. My parents know me better than to argue, and the new kittens were christened FM, CT and Peter.

The three grew up and it wasn't long before their personalities blossomed. CT was the cheeky boy, always larking about, Peter was the nutty one and FM was the quiet one. The only female cat, she was my favourite and would come and sleep on my bed. She was a prima donna, and when she wasn't loafing around the house like she was the lady of the manor, she was in the shed. The shed was her domain. It caught the sun, and she would sit in the doorway and bask in it like it was the entrance to her own little house.

I remember Bill coming into the house once with a look of absolute glee on his face. He beckoned us to come to the door with him, and there FM was again, sitting regally in the shed doorway. Looking up, I could see the dark shape of CT slinking over from the toilet roof onto the corrugated iron roof of the shed.

CT moved to the front of the roof so that he could peek over the edge. Having made sure that FM was there, he backed up a step or two and then jumped on the roof, making a tremendous *thump!* FM jumped a foot in the air and shot out of the doorway, then off across the garden. I swear, if cats could laugh, CT would have been in hysterics!

FM got him back, though – she always did. They were the best of friends and played with each other like that all the time.

Arja Hone
Wellington
New Zealand

Monkey business

'It is our sense of humour that sets us apart from animals.' I can't remember who wrote this, but he was way off the mark.

Many years ago I visited a little zoo in Mackay, where a number of rhesus monkeys were free to mingle with the visitors instead of being caged like most of the other animals.

At one point a man was standing with his back to an orangutan's cage to photograph his children with monkeys on their shoulders. The orangutan quietly came up behind him, reached its hand out through a channel it had previously dug under the wire of the cage, and grabbed the man by the ankle.

The result was exactly what you'd expect – one very alarmed man, and his wife and children similarly in shock. But the orangutan only held the man's ankle for a few seconds – just long enough to give him a good fright! Then it let go and went racing round the walls of its cage laughing its head off – obviously there's nothing like a good stir to relieve the boredom of sitting around doing nothing all day, even for an orangutan.

There was an interesting sequel to this. As I was approaching the exit gate I had a monkey on my shoulder and, despite there being

plenty of other monkeys available, a well-dressed older man who'd just entered the zoo with his wife and children grabbed the monkey from my shoulder and put it on his. I was stunned by his rudeness, but I didn't need to say anything to the man as the monkey said it all for me.

The man was standing facing his wife and children with an expression of total smugness on his face. This slowly dissolved into a look of utter disgust as wetness penetrated the shoulder pad of his suit-coat and he realised what had happened – the monkey had urinated on his shoulder! Perhaps this obnoxious man had saved me from having the same experience, but I think probably his abruptness had frightened the monkey. In either case, the monkey gave him a well-deserved lesson. I can still imagine the voices of his children from time to time in later years. 'Daddy, remember the day when the monkey peed on your shoulder?'

Peter Schaper
Biggenden, Queensland
Australia

Sometimes you need a little pampering …

When I was a kid growing up in England we had two really great dogs. One was a mongrel called Dandy and the other one, Jason, was a blue Great Dane. Dandy was a bit daft, but Jason was a clever chap. He did really neat things, like playing ball with Dandy by throwing the ball using his mouth for Dandy to bring back. He also used to carry Dandy's dinner bowl inside his own bowl (which was a large washing-up bowl) out to the yard, place the bowls next to each other so they could eat together, and then bring the bowls back inside when they were done.

Arthritis can be a big problem for Danes, so as Jason got older we all watched him for signs that he had it – especially Dad, who loved Jason dearly.

One day Jason couldn't get up from his blanket. We were all very sad. I was brought up in a pub and we lived over the bar. Dad got some of the men from the bar to come up and each take a corner of Jason's blanket so that we could at least take him outside into the sunshine while Dad called the vet.

Dad and the men struggled down the narrow stairs, with the horse-sized Jason lying heavily and awkwardly on his blanket. They got to the bottom corner of the stairs, and were figuring out how to turn the blanket to get Jason through the door when Jason calmly stepped off the blanket and wandered outside to pee!

He obviously felt like being pampered that day. I said he was clever!

Kirsty Buggins
Wellington
New Zealand

Patches got in first

I was out riding my beloved horse Patches one afternoon, when the sudden urgent need to wee took over any other thoughts floating around in my head at the time. In desperation, I looked towards the road near the track I was riding along. It was quiet, not a car to be heard – but just to be safe I headed into the bush. I feared that if I didn't hurry it was going to be a wet ride home!

I dismounted, found a well-hidden spot among the trees, and carefully and quickly made a nice clearing so that no leaves or twigs would poke me in any awkward places. Holding onto Patches I pulled

Patches: butter wouldn't melt in his mouth

down my jodhpurs, and was just about to position myself when I heard the sound of running water. Much to my dismay, Patches had already beaten me to it. I couldn't believe it; Patches never did a wee when we were out riding together – never.

Patches turned to me with this relieved, triumphant look on his face, as if he was saying, *Gee, thanks, Mum*. My nice clearing was no more; I would have to find a new one. Patches was laughing at me, I just knew it!

Patches was, without a doubt, the funniest, most humorous and most entertaining – not to mention cheekiest – horse I have ever had the pleasure of knowing. I miss him so much, but I am the luckiest person ever, for having been blessed with his love and companionship.

Karen Mol
Perth
Western Australia

Write to me ...
Karen Mol
7D Chailey Place
Balga
Perth WA 6061
Australia
or email Karen
misslestat@hotmail.com

A shih-tzu with a sense of humour

'Cheeky' was Oggie's middle name. The first time I saw him, in the pet department of a store in Sydney – a little grey and white fur-ball – he put his nose and paws through the bars of the cage, looked at me ... and won my heart.

He was a pedigree-free shih-tzu, big for his breed, long-nosed and absolutely gorgeous. He also had a wonderful sense of fun and an extensive vocabulary. His long life was a tale of memorable incidents which delighted everyone who knew him. I remember the time we came home to find every tissue in the tissue box spread across the floor, and a very proud puppy waiting for us. Then there was the episode of the escaped chicken, with Oggie in hot pursuit followed by a line of children, then a line of adults, all racing round the garden like a scene out of a farce. And, most remarkable of all, the occasions when Oggie would sing along to a saxophone solo on CD, crooning and howling softly, up and down the scale.

Whenever we were away from home Oggie was cared for by a friend, who enjoyed his company so much that she acquired a little Lhasa apso for herself. Unfortunately, Oggie had come to regard her home as an extension of his own and he looked upon the Lhasa puppy, Lama, as an interloper. Bossy as ever, he even had the gall to knock Lama off the bed and make him sleep on the floor for two nights, before condescending to let him back up again.

Sweet little Lama, only half Oggie's size, was unfazed by Oggie's grumpiness. He would simply challenge him to a wrestling match – a favoured pastime of Himalayan dogs – which Oggie found hard to resist. The sight of the two of them standing on their hind legs with their paws on each other's shoulders, wrestling away, was a pleasure to behold.

During one of Lama's visits, when the two dogs were racing round the park, Oggie had the bad luck to slip a disc. Under the vet's care he made a full recovery, but the next time Lama came to stay it wasn't long before he started manifesting all the symptoms of the old injury, the same yelps and whimpers of complaint. We took him to the vet. He hobbled all the way like a seriously crippled dog, only for the vet to assure us there was nothing wrong with him. Whereupon he trotted happily back home with his characteristic bouncy walk, completely cured! This pattern of behaviour was repeated every time Lama came to stay.

Oggie was never one to wait for an invitation. One evening, on a stroll along a quiet street of Victorian terraced houses, he was walking ahead of me when he suddenly took a sharp right turn through an open front gate, up the path and through the door, where a party was in full swing. By the time I caught up with him he was tucking into chicken and smoked salmon sandwiches that were being pressed upon him by the guests.

With an unusually deep bark for a shih-tzu he was an excellent guard dog, though his mode of expression had a certain style. Our home was on a corner block abutting the pavement, with the garden gate at the far end of the long side and a raised terrace on the corner. He would lie with his nose under the gate and bark furiously at any passing dog and owner, then race through the house and be waiting for them as they reached the terrace – at which point his bark was just a hand's breadth from the human's ear. To add insult to injury, he would then cross the terrace to accost them again as they rounded the corner.

There were risks involved with this behaviour – which he understood only too well, because he knew where many of his victims lived. Often, when taken on a local walk, there were streets he flatly refused to enter. On the odd occasion when he did get a taste of his own medicine, he would trot past, head high, feigning deafness – because, of course, it isn't proper canine etiquette for the passing dog to backchat the dog in residence.

Yet Oggie wasn't always hampered by his own rules. His favourite walk was down the back lane, which had a long row of rear garden gates, each with a dog's nose poking through the gap. He would

rise supremely to the occasion, prancing and rushing at the gates, pirouetting like a ballet dancer and thoroughly showing his disdain for the dogs locked behind them. It was wondrous to behold, but not something we could let him do too often because the ensuing racket was amazing.

The Great Barking Match was another memorable occasion. Some friends had acquired a black giant schnauzer named Bird. Oggie had never seen such a big dog before and voiced his amazement in a spate of barking, while Bird lowered his huge head to get a good look at this pint-sized interloper. Suddenly he let out a bass *WOOF!* of tremendous volume, which sent Oggie reeling back several paces. But this was great fun, and a lengthy contest ensued as they alternated single barks – *Woof! WOOF!! Woof! WOOF!!* – while all the humans in the room doubled up with mirth.

Oggie would also hold widdling matches with other dogs, each dog cocking his leg on the same spot alternately, the winner being he who cocked his leg last. Here again, Oggie had style. He used to hurl his leg up like a dancer, and once hurled it so far he overbalanced and fell over.

There was really no limit to Oggie's cheek. One hot Sydney afternoon he accompanied his master to an inter-office cricket match. Halfway through the game the sky darkened and the heavens opened. His master rushed for the car, parked about a hundred yards away – but did faithful Oggie run with him? Not a bit of it. He stepped into the shelter of the nearest Moreton Bay fig tree and waited for the car to come and pick him up. A hurtling ball of wet fur shot into the car the moment the door was opened.

When Oggie was 12 we moved to England. He was happy enough in quarantine with frequent visits from his friends, but the day we brought him out he gave us very moving evidence of his emotions.

Oggie had a long life of memorable incidents

Walking home through the wood, he made to bark but all that came out was a squeak. His happiness was so intense that he had lost his voice, and it took him six weeks to recover it.

He lived for five more happy years, as cheeky as he had ever been. The seal on his career came when he was one of four dogs who happened to be with their owners on the spot when Princess Anne arrived by helicopter on an official visit to Winchester. Oggie had long since got his bark back and, together with the other dogs, he used it.

The Princess was her usual gracious self. As her car passed, she leaned out and waved at him and his companions with a broad smile!

Patricia Johnson
Brentwood, Essex
England

Write to me … ✉
Email Patricia
strophalos@aol.com

Cocky's story

In the 1940s it was a rare occurrence for a sulphur-crested cockatoo to be flying around East Brunswick in Melbourne, Victoria. One day, upon hearing screeching, my grandmother (whom we called Noni) went out to investigate. As she looked up at the sky the glare of the sun made her place her arm across her forehead to shield her eyes. All of a sudden the shadow of a large bird got rapidly bigger as Cocky flew down and landed on her arm and face!

After the initial shock and excitement, Noni took Cocky inside to contemplate what to do next. Cocky was obviously a tame bird and spent some weeks calling out, *Yoo hoo, Mrs Ackland, it's*

Mrs Jones! Noni placed many advertisements in the 'Lost and Found' and in shop windows in search of a Mrs Ackland who may have had a friend or neighbour called Mrs Jones, but all attempts to find her proved fruitless. So Cocky had found a new home!

Cocky had a special cage in the kitchen at Noni's house, where he rapidly learned to mimic all her utterances and phrases and the way she coughed and sneezed. Many years later, after Noni had passed away, it was like she was still with us as every day Cocky would interject and contribute to conversations just as she would.

Cocky's wings were never clipped and he always had the freedom to come and go as he chose. In the mornings Noni would take him outside to the big old fig tree, which he would climb and then fly off and explore around the neighbourhood, always returning before dark and calling out to Noni to let him back inside for the night.

The family estimated that Cocky was not a young bird when he 'flew in'. He was quite a scruffy-looking thing and was thought to be around 30 or 40 years old. The longevity of these birds was made evident when Cocky grew up with my mother and uncle and then came to live with our family, surviving my childhood and adolescence and young adulthood. We figured 40 plus 30 plus 25 made him at least 95 years old.

Cocky was boss of the family pets, insisting that the dogs chauffeur him around the house on their backs. If any kids had their bikes out, he was there on the handlebars leaning into the wind – the faster the better. Cocky saw it as his responsibility to yell out to the neighbourhood kids if they were playing cricket on the road to *Get off the road, you'll get run over* or *You're out! You're out!*

He eventually lost his ability to fly, as he slowly lost his feathers and was left bald under his wings. He always retained his crest feathers

and the feathers on his wings and tail, but his body was buck naked! Even without the ability to fly, however, Cocky was never idle. He would turn his perches into toothpicks in no time at all and he had numerous toys he would play with.

Cocky also loved to dig, and some holes were so deep he would disappear into them with only his tail feathers poking out. He would talk to himself constantly while in these holes, sometimes whispering, laughing and singing, appearing to delight in his articulation of the spoken word.

My dad and Cocky had a love-hate relationship which occasionally resulted in Dad getting bitten. One time Dad was washing the car on the roadside and Cocky was playing and showering in the water run-off, when a car suddenly sped into the street. Dad gently guided Cocky to safety under our stationary car with his bare foot, resulting in Cocky chomping into Dad's big toe, thinking he was going to be stepped on.

Cocky had Dad so intimidated that once when Dad was on a ladder cleaning out the gutters we heard him calling out for help. We all rushed out, fearing he had fallen, but to Dad's embarrassment and our great amusement we found that Cocky had silently climbed up the ladder behind him and had Dad cornered. Cocky would not let him climb down the ladder and was trying to bite Dad's feet whenever he tried.

Cocky found many ways to torment Dad but the icing on the cake was when Dad, who was a bricklayer, was bricking up a shed in our backyard. After a long hard day's work on this task Dad was horrified to find that Cocky had been quietly at work removing all of the mortar from between the bricks from the bottom two rows of the shed wall!

Julie with Cocky baring his feather-free belly in 1985

Cocky's timing was impeccable and could be embarrassing for us all. On one occasion he waited until Mum was on the phone to a plumber about repairing some pipes. He let Mum introduce herself, then proceeded to ask in his female voice, *Who is Mummy's boy then? Come on, give us a kiss!* Mum was wondering why the plumber had faltered and then fallen silent, until she realised what Cocky was yelling out and then, very embarrassed, had to explain it was our cockatoo.

Every memory of Cocky is a funny story. He made such an impact on three generations and he was indeed a much loved member of our family.

Julie Raverty
Echuca, Victoria
Australia

Fun at the fair

A few years ago my boyfriend and I travelled around New Zealand working on the sideshows at the A & P Show (an agricultural fair or carnival). One day while in Masterton we found a tiny ginger kitten with the bluest eyes you have ever seen. He seemed very lost, so I took him home, fed him and fell in love with him. We called him Karni.

Karni travelled with us everywhere, to almost every town between Auckland and Wellington, and even came to live with us in Christchurch for six months. He loved an adventure. When he was small he would come to the supermarket with me and sit on my shoulder. He loved being wherever the action was when we were setting up the show. His best friend was a Dobermann. They would curl up together and go to sleep.

One day we were watching some horses showjumping. Then we saw Karni. He would wait until the horse was just about to make the jump, then run out in front of the horse and try to scare it – the showjumping people were not impressed! Poor Karni, we had to lock him in the car.

Then he took to stealing small toys out of the games at night. He had a favourite kind of toy – it was small, fluffy and squeaked. One morning we woke up and found 25 of these toys on the ground outside our caravan. He had brought them back to us one by one and they all said 'I love you' on them.

He took a dislike to one of the game owners and would purposely go into his game stall at night and knock all the toys off the shelves! He was very clever and everyone loved him.

One day, when we were in Auckland for the Easter Show and were staying in Manurewa, Karni disappeared. I was so sad that I cried for days. I knew I would never see him again but hoped that maybe he had found another family that loved him as much as we did. I will never forget our carnival cat – he was the best.

Joy Hesp
Foxton
New Zealand

Decoy for a bushranger

My husband's family was once owned by two cats called Blackie and Misty. Misty was a sook but Blackie was made of sterner stuff, and the two of them devised a cheeky game making use of Misty's failing.

The intrepid Blackie would hide under the hydrangea bush at the side of the house, while Misty trotted out between the roses at the gate and along the footpath with his tail held enticingly high, pretending he was unaware of any danger.

As the local cats knew Misty was fair game, it was never long before one of them accepted the challenge and gave chase.

Misty could only just hold his own as he galloped for home, and he would skid round the gatepost in the post-and-rail fence with his prospective victim close on his tail. Then he would zoom down the driveway and past the hydrangea bush.

Blackie, like any good bushranger, picked his moment to perfection, flashing out from his hiding place in time to give Misty's pursuer a good dusting. Vanquished, the stranger would bolt as soon as he could free himself, leaving Misty and Blackie strutting around with big grins on their faces.

This trick was carried out time and again, and there was always some sucker to take the bait.

Wendy Willett
Russell Island, Queensland
Australia

The feathered Dobermann

His name was Arnie and what a character he was. This little ringneck parrot provided our family with many laughs and much entertainment. He was always full of mischief and for some reason took quite a fancy to me.

My every wish was his command and this was to cause much trouble in my household. Arnie listened to my every word with an expression of complete comprehension. He was my little sidekick – but, thinking back, I was more his sidekick than he was mine.

Now, it came to pass that my little green 'monster' developed a string of rather clever antics, which could be executed on command. I could quite literally be sitting anywhere in the house with Arnie and tell him that I was very 'upset' with one of my family members, and he would have the greatest pleasure locating that specific family member and initiating an attack. This soon became a fun game: Arnie and me versus my sister. It always ended up with my sister hiding under her duvet (quilt) and begging me to call off my 'feathered Dobermann'.

Another trick which we loved to display to one and all was called 'One Leg, One Beak'. When I said these words he would put his beak to the floor and lift up one of his legs for me to hold. I would hold this leg and say again, 'One Leg, One Beak', which would start him moving like a little plough around the table. Very adorable, but even more so was the chuckle and conversation that followed and the big smooching noises. It was usually something cheeky but appropriate to the situation.

Arnie was a lovable and very affectionate parrot, but a cheeky so-and-so with tons of mischief to boot. A sorely missed member of our family!

Tracy Pitout
Auckland
New Zealand

Write to me ... ✉
Email Tracy
kpitout@gmail.com

Sure, animals are smart, but do you know a smart person?

Do you know a special person who will do just about anything to help animals – a true hero of the human species? Someone who has made a real difference to animals' lives? We'd love to hear about them.

The best submissions received will be published, with your name, in future SMARTER than JACK books. Animal lovers the world over will then find out how special this person is.

Submissions should be 100–200 words. For submission information please go to page 146.

4

Smart animals express themselves

It's too cold!

My friend Reiko has a very small toy poodle, sometimes referred to as a 'teacup poodle'. As Nemo weighs only two kilograms and the winter climate in Christchurch is quite cold, Nemo owns a selection of garments to wear outside.

One cold morning Reiko opened the outer door and told Nemo to go outside and pee. Nemo recoiled from the cold air and hurried back to the bedroom, where her clothes were stacked on a chair. She selected the warmest garment – a red alpaca sweater – and pawed at it, indicating to Reiko in no uncertain terms that she was not going outside until it had been put on her.

Colleen Borrie
Christchurch
New Zealand

He rules his kitty kingdom with an iron paw

My Horatio is a pretty smart cat. At 16 years he's still going strong, but he doesn't really like any company in his home – even my parents.

When my parents come to visit we often play cards. Horatio just can't stand this intrusion into his little world. He has never liked to

Horatio makes himself heard

share me, and wants to be the centre of attention at all times. He starts banging the kitchen cupboard doors. Next he'll jump up on the dishwasher, and rattle the wall calendar until it falls on the floor. Then he cries. Now, I don't know if you've heard an upset Siamese being vocal, but they can be really loud.

If I still don't pay him any mind, he rolls his food container between his paws (I keep a plastic container of special cat food as a treat on top of the dishwasher). If I still ignore him, he will knock the container onto the floor with a crash. I learned quite quickly not to use a glass container! Then he will knock the phone off the hook, so it lands on the tin food containers which store his normal food. Then there is more crying at an increased level of agitation.

Much to my parents' amusement, I've found that if I pull a high stool up beside me Horatio will come and sit regally upon it. He will sometimes watch the game, but mostly he sits and stares at me with love in his eyes. No more crying or carrying on. If I take too long getting the stool, he will sit on it with his back to me, glancing over once in a while to see if I notice that he's still *very* upset with me. His antics really amuse my folks.

He just doesn't want to miss out on anything, and most of all he just wants to be near me. How can I get mad at him for that?

Sharon Anne Serbin
Regina, Saskatchewan
Canada

Write to me ... ✉
Sharon Anne Serbin
68 Carmichael Road
Regina SK S4R 0C5
Canada

True understanding

Forrest is a German shepherd with a real chilled-out Kiwi temperament. He moves only if he has to or wants to. He knows many commands, but one he genuinely understands is 'Show me what you want'.

If he needs something, he will stare at me until he gets my attention. I then say, 'Show me what you want' and he will leap up and lead me to the door if he wants to go out, or to his toy if he wants to play, or to his food bin if he is hungry.

But the occasion which convinced me he completely understood the meaning of 'Show me what you want' was when he made me

follow him to the fridge, where he banged his head on the door three times so that I would open it and give him the remains of the juicy bone I had placed in there the day before.

Lesley
Waihi, Coromandel Peninsula
New Zealand

A pesky problem

Anyone who has been stung by a wasp knows just how painful it can be and is probably wary of these insects in the future. My hearing dog Roddy, a little papillon, has his own method of dealing with them.

Roddy was donated to Hearing Dogs for Deaf People when he was four months old. After being socialised with one of the charity's volunteers, he came into the training centre in Buckinghamshire for his advanced sound-work training. As part of this training he was taught to respond to everyday household sounds, like the doorbell and telephone, by touching his trainer to alert her and then leading her to the source of the sound.

Towards the end of his training Roddy was unfortunately stung by a wasp, and had a very severe allergic reaction which resulted in him collapsing and being rushed to the vet for emergency treatment to save his life. It was touch and go for a while, but Roddy recovered and successfully completed his training, and then came to live with me.

One day, not long after he moved in, Roddy rushed over and started scrabbling at me, which is his way of alerting me to something

Forrest knows exactly what he wants

going on. When I asked him, 'What is it?', Roddy led me back to his dog bed and I immediately saw there was a wasp buzzing round his bedding. As soon as I had disposed of the insect, Roddy happily curled up in his bed and went to sleep!

Mrs D Amos
Somerset
England

You can't hide from me!

In my grandparents' neighbourhood there was a semi-stray black cat called Skitz whose owners were never home. She started to hang around with my grandparents' elderly cat and they ended up adopting her.

Skitz is forever losing her magnetised collar so she doesn't always have a magnet to open the cat door at feeding time. So she pats on the door with her paw and Grandad comes running to let her in.

However, this particular day, Grandma and Grandad were at their next door neighbours' house. Skitz must have been to the cat flap and, after realising they weren't home, figured that they were still next door.

Grandma and Grandad were sitting in the neighbours' lounge having a good old chat when they heard something patting on the window. They looked up to see Skitz peering in as if to say, *You can't hide from me. Now where's my dinner?*

Sarah Bull
Kapiti Coast
New Zealand

A naughty runaway

Sint Maarten is a small island – a total of 36 square miles. Many people come here on contract to work for three years or so and then return to their 'normal' lives. Of course their families come too, and they quite often want a dog.

We were given Max by a family who was returning to Holland. They were at their wits' end because he would always run away. Open the gate, he'd run away – call him back, he'd run away – chase him, he'd run faster – let the children take him for a walk with a leash, he'd run away. The call from the local vet was always dreaded: 'Come and get your dog, some people have handed him in here.'

So we got Max knowing he was a 'bad' runaway golden retriever. And, sure enough, he ran away. The second you opened the gate or turned your back on him while out for a walk or left him in the garden, he'd find a way out and go adventuring. (We subsequently found out that he quickly trained a neighbour to open the gate for him when she got home in the morning.) He'd keep his head down, never meet your eye and just get on with his life. For a couple of weeks we tried all the normal things: scolding him when he returned, withholding his meals if he wasn't there at mealtimes, catching him and taking him straight home …

Then one day when he came back from an adventure, pleased as punch, he greeted every one of us and we each turned away, ignoring him. He was sad and confused and tried the greeting round again with the same results. Next morning when everything was back to normal he was quite happy again. Upon his return each time he ran away, he was not greeted but completely ignored, and excluded from the hugs and cuddles the other dogs received and any games we played. The other two dogs were called inside for sleeping and, if

he didn't make it through the door in time, he slept in the garden by himself – even if there was a thunderstorm. Next morning everything would be back to normal. After a week of this treatment, he quit running away and would wait politely to be taken for a walk. We congratulated ourselves on a job well done.

My husband Andrew had to go to a neighbouring island and was away for the night. The next day he came back at lunchtime. The two bitches jumped up and greeted him and were greeted in return. Max lay down with his back to Andrew. If he happened to catch Andrew's eye, Max would turn his head with exaggerated nonchalance and look in another direction or simply close his eyes. This made us quite worried, as Max generally has a poor constitution and we were concerned that perhaps he had eaten something bad. We'd just lost another golden retriever to cancer so we were somewhat sensitive to nuances of behaviour.

That evening Andrew tried feeding the dogs so he could check on Max. But Max didn't appear to see the bowl of food, nor Andrew pointing to it. He just sat quietly until Andrew called me outside to observe this strange behaviour. When I greeted Max and told him he was a good boy and could eat he tucked right in – and then we both got the message: Andrew was a naughty runaway owner!

Lyn Rapley
Sint Maarten, Netherlands Antilles
Caribbean

Florentino and Parvati check out what's on TV

Television critics

My friend Erica often came round in the evening to watch television with me.

Florentino, my red tabby cat, and Parvati, my black cat, usually sat on our knees – Tino on mine and Vati on Erica's – and watched TV too. If anything of particular interest to them was on they hopped down and sat about three feet from the TV, watching it intently.

One night a programme showed footage of a disaster in Turkey. Suddenly we saw an earthquake-ravaged zoo: walls collapsed, distraught animals confused and milling about, making a cacophony of sound as their voices united in terror.

Tino jumped down from my knee and ran to the TV. He stood on his back legs and hit the edges of the screen dials with his front paw until he managed to turn it off. He then returned to sit contentedly on my knee.

Beverley Brown
Paraparaumu, Wellington
New Zealand

Write to me ...
Beverley Brown
Summerset
92-65 Guildford Drive
Paraparaumu
Wellington
New Zealand

Training needed

My grandmother had recently passed away, leaving behind Chow, a silky terrier cross.

A few days later, the family was sitting round Gran's kitchen table planning the funeral etc. Gran had always fed Chow at around 5 pm and her dinnertime had well and truly passed. Unfortunately nobody had noticed how late it was getting.

Chow realised that we were not taking the hint that she was hungry, and quietly disappeared for a little while. As her bowls, basket, etc had been relocated to her new home, we had fed her the previous day in a disposable plastic bowl. She went to another part of the house and retrieved this bowl.

We were all surprised to see her come walking back into the kitchen carrying the bowl in her mouth. This was quite out of character for Chow – she had never done anything like that before.

Delightful Chow

Obviously she must have realised that her new owners needed training!

Stacey Ebben
Karuah, New South Wales
Australia

Write to me …
Email Stacey
ssburben@bigpond.net.au

His message was clear

My son owns a fox terrier – a very intelligent creature.

The other day when my son's wife was hanging out the family's washing, the dog stood with his head to one side, watching her very

quizzically. All of a sudden he ran off to his kennel, and returned dragging the rug that he sleeps on and dropped it at her feet. She started scolding him and told him to take the rug back. But when she picked it up she found it was soaking wet, and so hung it on the line.

Once the dog saw that she'd got the message, he raced back to his kennel and came back with his empty drinking bowl in his mouth. He put it under the garden tap, which is where they usually filled it. He then sat back and looked up at her as if to say, *Can you fill this up? As you can see it got knocked over and wet my rug!*

Rose Field
Richmond, Nelson
New Zealand

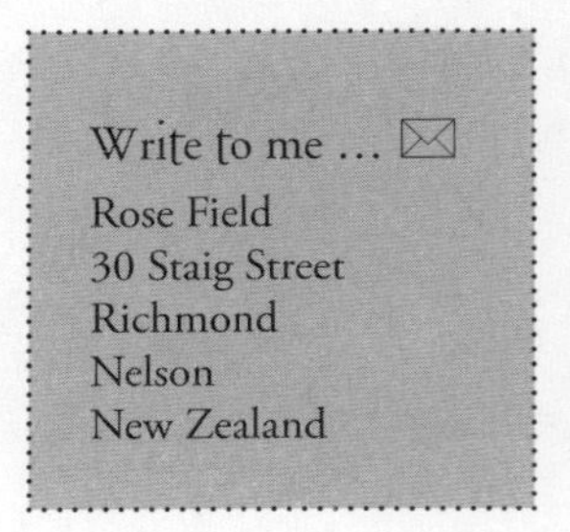

A musical dog

This is a story about a musically discerning little dog named Josephine.

Years ago, I was given an ocarina, a little clay flute with six holes. However, it definitely doesn't play six notes – in fact, it plays more like three, making it a rather tuneless little instrument. I very rarely play it, and I certainly haven't mastered it. So when I picked it up for the first time since Josephine had joined our family, it wasn't long before my husband called out, begging me to stop. Josephine, however, just stood and stared at me as I played, tilting her head and moving her ears around as she grew accustomed to the new sounds.

Later on, my husband and I were sitting on the sofa. My husband started to play the guitar – just the few chords and snippets of songs he knew. Josephine hopped onto the sofa to join us, and seemed quite content to listen to my husband's musical interpretations.

Then I had the bright idea that we could play a 'duet' – my husband on the guitar and me on the ocarina. So we began. Unfortunately, with my ocarina's limited range and my husband's limited chords, it was little more than a raucous noise.

And what did Josephine think of all this? Well, she voted with her feet: she jumped off the sofa and trotted off to the bedroom. Twice my husband called her back, and each time she obediently returned. But then, as soon as we started our little ensemble, she would trot off again to the sanctuary of the bedroom.

Clearly, Josephine's musical sensibilities could not withstand such distasteful harmonies!

Olivia Thatcher
Auckland
New Zealand

Marguerite's change of heart

We lived on a dairy farm for a while. We bought chooks and dogs and I said jokingly, 'All we need is a goat.' My husband took this as a hint! He bought a goat for me on our second wedding anniversary.

Marguerite was a Saanen goat – large and white with greenish eyes. She took one look at me and it was instant mutual dislike.

She had to be milked, but Pieter was the only one allowed near her. Marguerite would look at me, her green eyes glinting, and bleat, *Naaa, naaa, naaa* – goat talk for *Ha! Look who's got him now!* She'd

wait until the seedlings were just about to flower, then she'd bite them off. She'd wait until I was away and then she'd eat the clothes off the line. We'd end up with little pegged fringes.

'Let's get rid of her,' I'd say. 'She's more trouble than she's worth.' Until one day.

It was mid-morning, the cows had been put in the house paddock and Marguerite was in there too. The cows started milling around like a big bovine whirlpool. Then, as one, they turned on Marguerite and started butting her. Marguerite started bleating. The cows inched forward, battering her all the time.

I ran out into the paddock with a broom in my hand and began hitting their rumps. 'Get away! Get out! Leave her alone!' I screamed. The cows lumbered away, suddenly looking bored. Marguerite was quivering and making small bleating sounds.

'Come on, Marguerite, you're all right.' I took her to the shed and she lay down on some burlap bags.

From that day on, Marguerite was my shadow. She followed me wherever I went and never ate the seedlings or the washing again.

Ruth Boschma
Mont Albert, Victoria
Australia

Write to me ...
Email Ruth
ruthboschma@hotkey.net.au

Here's a cheeky photo: a bold cat meets a curious bull

Help shape the future of our books

We would love to get your feedback on new ideas that we have for the SMARTER than JACK books. We're often coming up with concepts that are sometimes wacky, sometimes not. We need your help to work out what we should and should not do!

A short questionnaire may be emailed to you when we have a big decision to make. All responses will be kept confidential. Your input will help build the future of SMARTER than JACK.

To join our test group go to www.smarterthanjack.com/mainsite/concept-testers.html.

Your say . . .

Here at SMARTER than JACK we love reading the mail we receive from people who have been involved with our books. This mail includes letters from both contributors and readers. We thought we would share with you excerpts from some of the letters that really touched our hearts.

'I got a SMARTER than JACK book for my birthday and I have to say it's a great book which shows people that humans may be smart but animals can outwit us anyway.'

Kelsey, Canada

'I wanted to say thanks for the "Heroic animals are SMARTER than JACK" book. I just opened it and had to read a couple of stories. I read the one on page 8, "Our precious jewel", and it made me cry!

I will be taking it with me this weekend to read and I know my mum who is visiting me next weekend will be dying to read it. (She loved the first ones!)'

Elise, Australia

'Thank you for the copy of "Animals are SMARTER than JACK". I appreciate it, and so will the person I intend to give it to, my big sister Anne – she was Donna's original owner when she was a puppy. The photo came out really well, and I know it will bring a tear to Anne's eye as it did to mine. Thank you for the reminder of Donna.'

Tony, New Zealand

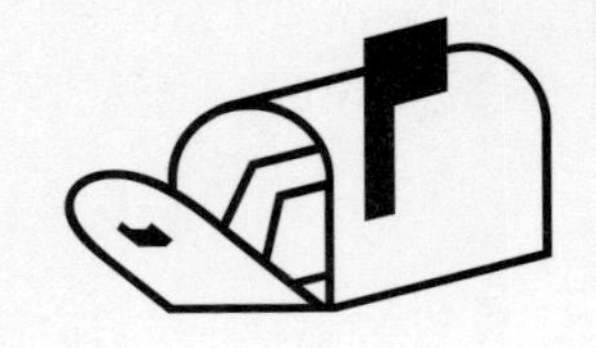

'Thank you for my "Animals are SMARTER than JACK" book. I love it, though I did stay up till midnight to read it all – not such a bright idea.

I will certainly be looking closer at animals in the future.

I was amazed at the story about the sheep, as I grew up on a sheep property and have always thought sheep were the dumbest creatures on earth. Well, looks like I was wrong, as Adrian Holloway pointed out in "Lamb care centre".

Ivodell, Australia

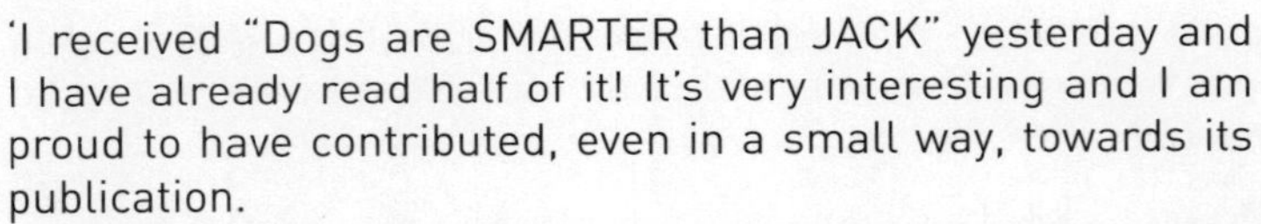

'I received "Dogs are SMARTER than JACK" yesterday and I have already read half of it! It's very interesting and I am proud to have contributed, even in a small way, towards its publication.

I have been promoting it and I really hope it will raise a lot for our little friends.

I am already thinking of my next story!

Thank you for your help, and good luck.'

Franca, England

'I received the books today, many thanks. Now for a good read and plenty of laughs.

My grandchildren will be visiting this weekend so they will enjoy the stories too. Their two dogs probably will also, as they usually sit with us and listen during story time.'

Diane, New Zealand

'Love the books, laughing one minute, howling the next, but the warmth ... more please!'

Maureen, Australia

'Thank you for making me laugh, smile and cry.'

Alumita, New Zealand

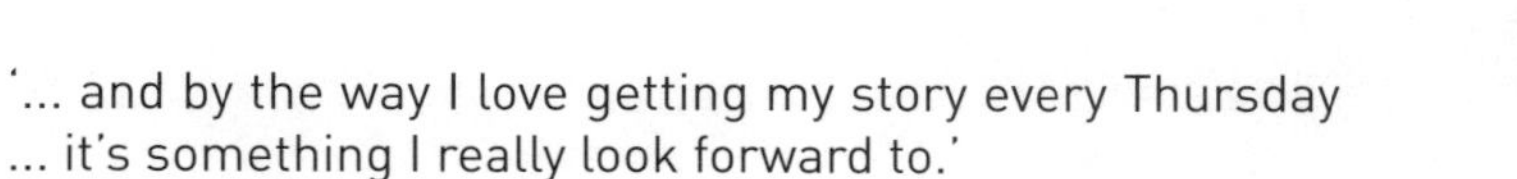

'... and by the way I love getting my story every Thursday ... it's something I really look forward to.'

Cathy, Canada

'Your books have taken me to a new level of respect and patience with animals, particularly with spiders and smaller creatures.

It's great to know that so much money has been raised for the RSPCA and overseas welfare organisations.'

Georgina, Australia

Lisa, Anthea and Hannah of SMARTER than JACK selecting stories

'Thank you – keep up the great work, I look forward to the stories each Friday.'

Mandy, New Zealand

'I've just collected my book with Dusters' story in it!

I walked back to work from the post office with a grin from ear to ear!

Thanks again for the opportunity, keep up the great work.'

Shannon, Australia

'Thank you so much for the copy of the latest SMARTER than JACK book with my story in. I am quite chuffed!

I've sent out heaps of emails about it, with your site, and do hope many purchase a copy.

It is a lovely book to be a part of. What a wonderful concept it is for fundraising.'

Margaret, New Zealand

'I got a "Dogs are SMARTER than JACK" book as part of a Christmas present (as I love animals but especially dogs) and I love it.

Lots of other people have picked it up and started reading it and love it too!'

Emma, Australia

5

Smart animals have fun

Birds at play

Thinking about animals having fun reminds me of two instances where I watched birds playing.

The first took place over a pond at Alder Place in South Surrey, British Columbia, where we lived at the time. A young sparrow had a small feather, which she held in her mouth while flying high over the pond. She'd let it go and then fly quickly down to catch it, over and over again as it neared the pond. When the feather (and bird) got quite close to the pond, the little bird would catch her feather and fly upward again to start the game over.

The second instance of birds playing that I saw took place in the ocean off Stanley Park, Vancouver. The tide was coming in and the water running swiftly. The seagulls lined up in the air over the ocean, coming down one at a time to 'ride the waves'. At the end of the ride, they'd fly upward again and join the queue in the sky. This went on for a long time. They were really enjoying their playtime!

ShirleyMae Shindler
Vancouver, British Columbia
Canada

Ready, set, go!

Rustler was a stray dog who followed my grandfather home one day and took up residence with us.

He was very much his own dog. He did not eat the food we gave him, preferring to go out and fend for himself, but he took his food and gave it to other dogs, notably his friend the Scottie dog who lived down the street.

These were the days before dogs were required to be confined to their owners' properties. We lived in what was then a village, with lots of fields and open spaces. We did not worry when our dog went out. Rustler and his friend used to wander off each day and return in the evening to their respective homes.

Rustler got his name because he was a born thief. He used to wait at the bus stop at the bottom of our street and jump up at people alighting from the bus, causing them to drop their parcels, which he would then bring home. My mother spent a considerable amount of time going round the village returning the stolen goods.

Rustler's most remarkable exploit was witnessed by my brother as he was walking up the lane after work. He saw Rustler with about half a dozen dogs all lined up in a row. At a bark from Rustler the dogs raced off for several yards, then returned to the starting point, where they lined up again and waited for the signal for the next race!

Rustler eventually left us, having decided it was time to visit fresh fields. His friend the Scottie dog disappeared at the same time, so who knows what adventures they had roaming the countryside together.

Kathleen Smeaton
Leeds, West Yorkshire
England

Catch me if you can

Quite a few years ago I went to a holiday site at a place near Plymouth in Devon, England. It consisted of log cabins built into the sides of hills rising about 30 feet above a large woodland glade. Animals were plentiful in the glade and easy to watch from the large windows of our cabin.

In the mornings we had our breakfast sitting at the window, and watched rabbits chasing each other around in the glade. Every morning a gang of crows also came to the glade. After a couple of days we realised that when the crows settled on the grass near the rabbits it was time for a big game of 'catch-me-if-you-can'.

A rabbit would chase a crow and, when it caught up with it, would give it a tap with its nose, whereupon the crow would turn round and chase the rabbit until it caught it to give it a tap with its beak, then run away with the rabbit in pursuit. This would go on for about 30 minutes. When the game was over the crows would take off and go about their business, while the rabbits went back to theirs.

The next morning the crows would arrive and the game would start again. We stayed there for two weeks and this took place every morning. There was only fun, no animosity, and I feel lucky to have witnessed it.

Joyce Taylor
Bristol
England

Mischievous Bella

Bella has a good laugh

I was dog-sitting Bella, a rather large mischievous black poodle with a beautiful nature, for a week while her owners were on holiday.

One day I heard a car horn beeping repeatedly. I ignored it for a while, thinking it was a neighbour, but it persisted. Then I realised that the beeping was coming from our basement garage, and thought it was probably my husband who had come home and wanted help up the internal stairs with some parcels. I raced down the stairs. But no! My husband was not there and I started to wonder what the heck was going on.

The horn beeped again. I looked at my car – and was flabbergasted, to say the least. There was Bella, sitting in the driver's seat with both paws on the steering wheel and tooting the horn, with such a look of sheer excited enjoyment on her face. Her expression was the closest to that of a dog laughing that I had ever seen.

My car at the time was a small Toyota Corolla. Just how did Bella – large dog that she was – squeeze through the half-open driver's window into the car without a scratch to the paintwork? I think our Bella had been at it before!

June Spragg
Warkworth
New Zealand

Write to me ...
June Spragg
453 West Coast Road
Ahuroa
R D 1
Warkworth 1241
New Zealand

A special bond

Our son has a very obedient black Labrador called Zara, whom his little daughter Molly treats as an almost human companion.

He has erected a wooden platform with a slide attached to it for Molly to play on in the back garden. At two and a half years old, Molly decided, all on her own, to teach Zara to go down the slide.

She climbed the steps and stopped at the top to call Zara, who climbed up after her. Then, commanding the dog to sit at the top of the slide, Molly slid down first. Standing on the ground below, she called, 'Zara, come!' Zara then stepped onto the slide and slid down to her.

We watched, amazed at both Zara's willingness to perform this feat on the orders of a tiny child and the obvious trust between them.

Judith Baxter
Wellington
New Zealand

Anything you can do, I can do better

When my children were small and my beautiful, darling Labrador/kelpie cross Winnie was young, we liked to go on what we called 'neighbourhood treks'. This meant walking to Mutch Park in the next suburb instead of the park across the street from our house. I'd strap Sean on my back, put the lead on Win, hold Drew's hand and we'd head off.

Along the way we talked about the trees, plants and flowers that we saw, and laughed at Winnie barking at a porcelain statue of a cat. She seemed to find it insulting that the cat would not run away.

One day at Mutch Park, Drew decided to roll down the big grass hill, so we climbed to the top and found a good spot. The two boys lay on their backs and rolled over the edge, turning over and over until they reached the bottom. They laughed like mad and ran back up to the top to do it again.

Winnie and I were at the top of the hill, watching them from under a tree. After the boys had rolled down a few times, Winnie got up and walked over to where they were preparing to roll again. After both boys had launched themselves down the slope, Winnie lay down on her side at the precipice.

I stood up to watch, wondering what she was doing. I saw her tuck all four legs in as close to her body as possible, and could not believe my eyes when she rolled over the edge. Over and over she went in a 'kneel-roll, kneel-roll' kind of action.

The boys had reached the bottom in time to see this stunt and cheered her on loudly. When she landed at the bottom she jumped up and ran to the boys, leaping around them with obvious joy.

This amazing scene was subsequently repeated many times, before other activities took our fancy. I still laugh at the memory of that big dog rolling awkwardly down the hill and being so very proud of her achievement.

Trish Haywood
Sydney, New South Wales
Australia

Write to me ...
Email Trish
thewilddingo@iprimus.com.au

A sweet gift

My mouse Jane was around six weeks old when she had her first litter. She gave birth to only one baby – Clair – who is now happily living alone in her very own cage.

One morning I decided to mix some Cheerios breakfast cereal in with the container of mouse food. When I tried some on Clair, she ran right up to a rice Cheerio and ate the whole thing up. After she'd realised how lovely they were, she went wild! First she strung a few, much like beads, onto the metal tube of her water bottle. Then she cutely furnished her home all over with the sweet cereal.

After she got bored with her new decorations, she picked one up, looped it over her tail and ran round the cage. If the Cheerio happened to fall off, she would punish it with a little nibble as if it had disagreed with her, until it was eventually eaten up.

I went to pick her up the next morning, and all but a few Cheerios had been eaten. Clair is very tame, and she ran up to me with a corn Cheerio – her second favourite type – and handed it to me. I actually still have it. It's my little gift, and it's in a little box along with some pellet bedding that she's given to me.

I suppose it just goes to show what can happen with a bored mouse and a few tasty bits of cereal!

Nora Fenn
York
Western Australia

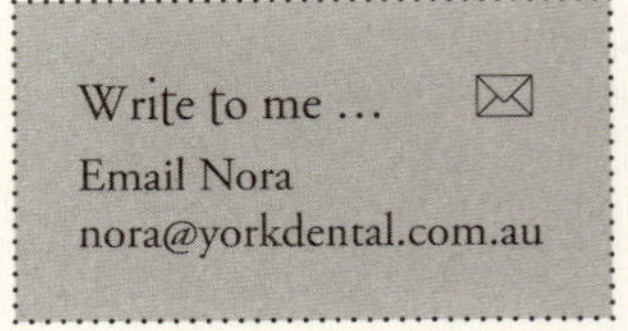
Write to me ...
Email Nora
nora@yorkdental.com.au

Rainy day fun

It was a rainy day so my father was jogging indoors, making clunky laps of the Victorian rooms in rolled-up pyjamas and floppy leather slippers. I had brought Cokee, my three-year-old Siamese, round to stay with my parents.

Cokee disappeared while Mother and I sought a collision-free refuge in a corner of the kitchen to chat. Periodically I could hear a 'Yip' from my father. After the third 'Yip', I finally asked, 'What is wrong with that man?'

'Oh, it's nothing. Cokee is helping Daddy run. He hides in a different place for each lap, then tags Daddy's bare ankles.'

MaryGrace McDermott
West Linn, Oregon
United States of America

Chummy: king of thieves

Chummy's story goes back many years to when I was newly married and bought him for one shilling from a pet shop. He was the last, lonely pup of a litter that had been on display in the pet shop window.

A pup of no particular breed, he proved to be a faithful friend with only one failing: he was a master thief! It was quite funny because, although he was prepared to steal anything he could, he did not approve of the cat taking anything. He created such a fuss that any attempt at thievery by Puss was quickly foiled.

It was during a time of rationing and a shortage of dried fruit etc that he committed one of his most notable crimes. Back then, dogs tended to roam the streets freely. Chummy went out for his daily wander and came back with quite a large fruit cake. He struggled

over the fence tugging this cake behind him. It was obviously a celebration cake which someone had worked hard to acquire the ingredients for, so you can imagine my horror when I saw it. As quickly as I could, I got a spade and buried the evidence in the garden.

On another memorable occasion he came struggling back with – of all things – a baked dinner on an enamel plate. Obviously some of the dinner had fallen off on the way, but baked potatoes and vegetables were clinging onto the gravy. As he must have left a trail I expected to see an irate figure chasing him, but no one appeared. Again, it was quickly disposed of.

I imagine he thought he was bringing us treats, as he always looked so pleased with his loot. He also used to bring home an assortment of clothes such as socks and underwear. He lived until he was 14, but his thieving was curtailed during the last four years of his life because he was blind. However, he still loved life and coped well. He was a most lovable thief.

Joyce Taylor
Bristol
England

Write to me ...
Email Joyce
pegotty2000@tiscali.co.uk

A firm favourite

This is a true account about a stray dog. I don't expect many people to believe me – I suppose that is why I seldom tell it. How I acquired Pooch (that was the name I gave him), or rather how he acquired me, is another story.

After the war, a fellow ex-serviceman and I and our families started a boating business on the Thames, about two miles upriver from Windsor.

Pooch had settled in and would travel with us on the bus to Windsor. Dogs were allowed to travel upstairs on buses so we became regular upstairs passengers. Pooch's bright friendly character soon made him a firm favourite with the conductresses. He began to board the buses by himself. The bus, which took the river road, stopped just outside our gate on the half-hour. The other stop, on the hour, which took the inland route, was just around the corner.

Pooch would wait at the correct stop; even if there were no other passengers waiting, the bus would stop and pick him up. He would go upstairs, and alight when the bus reached Windsor.

I wondered why he went to Windsor, and later I found out. I was in town with Pooch doing a little shopping when I went into a general store to buy some pipe tobacco. The lady behind the counter saw Pooch and told me he came in regularly and she would give him a bar of chocolate. Like the transport staff, she adored him.

Ronald G Moss
Buderim, Queensland
Australia

The smartest animal of all!

Many of our readers love to take photos of their pets reading (or sleeping) on SMARTER than JACK books – now, that is smart! We love getting these pictures and thought they should no longer be kept hidden from public view. In each edition of SMARTER than JACK we will publish the best new photo we have received.

For submission information please go to page 146.

This photo of Othello is by Maree Kirk of Wellington, New Zealand. Maree receives a complimentary copy of this book.

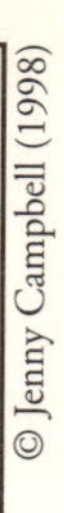

Here's a cheeky photo: curious Bix, a nine-year-old Jack Russell

6

Smart animals are creative

A pukeko in an apple tree

I love pukekos (also known as New Zealand swamp hens). My niece does not. To demonstrate why, she took me to the orchard, where one pukeko sat in an apple tree and eight pukekos waited beneath it.

We watched the pukeko up the tree snap off a big Granny Smith apple. It fell to the ground, and another bird snatched it up and whisked it away into the reeds.

The pukeko up the tree repeated this performance, snapping off one apple after another for the birds below. Eight apples were quickly taken until no birds remained on the ground, and the cheeky pukeko then dropped a final apple and swooped down to grab it, then waddled quickly into the reeds too.

How's that for know-how!

Kath Johnston
Kaitaia
New Zealand

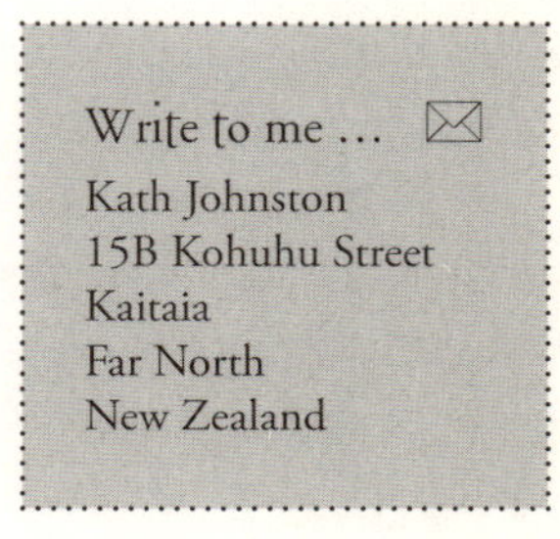
Write to me ...
Kath Johnston
15B Kohuhu Street
Kaitaia
Far North
New Zealand

A new way to play

One of our miniature poodles, Bonnie, is crazy about playing with tennis balls. In fact, some of our friends have tried to see how long it will take for her to get tired of running for the ball – and they got tired sooner than Bonnie!

The summer was very hot, so we bought a small inflatable pool for the dogs as they have always liked the water. I decided to show Bonnie that we could make her ball game even more exciting by throwing the ball into the pool so that she could then go into the water to get it.

She did this a few times, all wet and happy, but quickly realised that actually getting into the water to retrieve the ball wasn't necessary at all. She only had to step on the edge of the pool, making the water run out and the ball float towards her. Needless to say, the pool was very soon punctured by her sharp claws. But, before that, Bonnie had a few hours of good fun!

Katherine Shevelev
North Epping, New South Wales
Australia

An escapologist

I worked on a sheep farm in Mid Wales where there were also a small number of cows fattened for market.

One of my jobs was to let the calves out of a pen in the morning so that they could suckle from the cows. When the calves had finished suckling they went back into the pen at the side of the cowshed. I then let the cows out to the field and cleaned the shed.

One day as I was finishing putting down the fresh straw in the cowshed I felt a nudge at my arm, and turned round to find that all the calves had got out of their pen. I chased them back into it and bolted the gate, making sure the bolt was pushed home.

A few minutes later I discovered that they had got out again. The youngest ones thought it was a good game and went charging about. I rounded them up and returned them to their pen. I pushed the bolt across and made sure that the looped handle was pushed down as far as it would go.

I decided to stand back and watch, thinking that perhaps one of the farmer's sons was playing tricks on me. But imagine my surprise when I saw the biggest calf come up to the gate, sniff round and then put his tongue through to the bolt. After a few moments he had located the handle part and pulled it up with his long tongue. Then he nudged it sideways with his nose until the bolt came out of the slot in the gatepost. He pushed the gate forward with his nose until it swung open. All the other calves then followed him through the gate. I had to tie the gate firmly after that!

It wasn't long before the clever calf was sold on. I often wondered what happened to him.

Laurence Smith
Presteigne, Powys
Wales

Here's what you want

We have a five-year-old miniature poodle, Zak. One day my son and I took Zak to a local park where he loves to swim and retrieve.

The area around the water is grassy, with no objects to throw for a dog to fetch. After Zak had nagged and harassed my son for some time, my son said to him, 'Find me something to throw and I'll throw it for you.' Zak launched himself into the water and came

out with a very well-used thong (sandal)! We just about fell on the ground laughing, but Zak patiently sat there, thong in his mouth, as if to say, *What are you laughing at? I found something to throw!*

How did he know?

Maureen McComb
Montrose, Victoria
Australia

Write to me ...
Email Maureen
wiskas@iinet.net.au

Resourceful Zak

Sweet child o' mine

Well, I suppose everyone thinks their animal is very special – and they are. In my particular case, my beautiful domestic tortoiseshell female is outstanding in every way.

I got her on the spur of the moment from a friend of ours on a farm. A wild tomcat had managed to get to my friend's two females and ... well, you know the rest. I named my kitten Porsche.

Porsche soon figured out how to open the doors in the house we were living in. It was an old 1970s house, with high wooden doors and curved stainless steel door handles that were quite stiff and well worn. In the still of the night just as you were heading off to sleep, Porsche would jump onto the door handle and pull it down, then use her paw to push the door open.

Not long after that, she figured out how to unlock the cat door. We had recently moved house and noticed another cat coming in and stealing Porsche's food. My partner and I locked the cat door, so that no cats could come in but Porsche could head out when she needed to.

One morning before work we let Porsche outside. Upon returning, we noticed that Porsche was waiting for us in her usual place, inside the house by one of the windows. We wondered how she'd managed to get into a completely locked house. We checked all the windows, doors and the cat door. Everything was locked. We had our suspicions that she knew how to get in through the locked cat door.

Luckily, one day I was in the kitchen and noticed Porsche waiting outside her cat door, staring in. Although it felt a bit mean, I decided to wait and see how she got through that cat door.

Porsche wasn't coming through it, though – she'd much rather I opened the door and saved her the effort. So I pulled out my biggest weapon: her biscuits. My cat could be ten miles away and hear the

packet opening. I shook the packet. Soon you could see her claws unhooking the latch, then her paw sliding under and bringing up the flap so that she could climb in.

I would have thought this would be one of the hardest doors of all to open, although just recently Porsche has outsmarted me again by working out how to open the sliding doors in our rumpus room. She slides her claws between the two doors where they join together. She uses all her strength to pull these two wooden doors apart, then uses her head and pushes her body through the gap.

Porsche is a truly remarkable cat – one like no other. She has even managed to sneak onto buses at the bus depot (which used to be next door) and journey halfway round Auckland. The poor drivers had no idea Porsche was on their bus until she jumped on their knees and gave them a 'Where to next?' look. They realised to their dismay that Porsche had been on the bus since leaving the depot! Luckily the bus drivers got to know us, after many experiences with our cat travelling round on their buses. They would return her home after their day's work.

I have no need for a dog to keep watch over the house. As soon as there is a noise, whether it is in the still of the night or while the dawn is breaking, Porsche will stand up and head to the door.

Porsche is a friendly cat, to humans and to other animals of all kinds. She once brought home a lop-eared rabbit. Rest assured, the rabbit followed her home – it was not in her mouth. We even caught the whole thing on tape. Actually, to tell the truth, the rabbit chased her home. The rabbit would run after her and then she would run after the rabbit. Never before have I seen a cat playing with a rabbit in such a playful, innocent way, and seen animals that are usually enemies acting like best friends.

Porsche and I have a special 'smacking game' we play, where I will tap Porsche's paw lightly and she taps me back. We also play hide-and-seek. I run and hide and then she will come and find me, then she will hide and I will have to find her. If Porsche can't find me I will hear a faint meow, and then I will have to come out. We play the game over and over again.

Porsche is my best friend and companion. I consider her my child and I can't imagine my life without her.

Cindy Light
Auckland
New Zealand

A definite improvement

Devil was a magnificent black curly-coated retriever, and one of the most intelligent dogs who ever came into our lives. He was basically my husband Tony's dog, and they did most things together.

Devil's passion was our old Toyota Land Cruiser and it became his domain. He loved that smelly old Land Cruiser with its diesel, grease, and collection of general farm tools and rubbish. He would sit proudly on the tray out the back, king of all he surveyed. We always gave him the freedom of sleeping in the warm stable but, cold or not, his choice was always the Land Cruiser.

Devil's exploits were spectacular but one that stands out in my mind occurred several years ago. We had sold our property in Gloucester, New South Wales and relocated to Quirindi on the Liverpool Plains.

It was mid-winter and very cold and frosty. The Land Cruiser was left parked outside the house without any protection from the cold, with Devil as usual sitting on the tray out the back. Tony was working at the stable about two hundred metres away and can confirm that what happened next is true.

Devil jumped from the Land Cruiser and trotted over to the stable where some old carpet had been dumped. He grabbed the heavy carpet, dragged it over the rough gravel to the vehicle and proceeded to jump into the back of the vehicle with the carpet, which he pushed flat on the floor. Satisfied that the carpet was what he wanted, he sat down on it and eventually curled up and went to sleep – what an improvement! This was so much better than the freezing cold steel tray on the back of the Land Cruiser.

It is with affection that we will always remember him. Devil died two years ago and is sadly missed by us all.

Ann Easton
Quirindi, New South Wales
Australia

Get up!

We have two dogs and a three-year-old cat, Gypsy. One of our dogs, Kooch, is a 14-year-old ridgeback cross who is developing incontinence problems. Kooch had never lived with a cat before we got Gypsy, and when Gypsy strolls past her she often whines and whinges and gets up from her spot.

We keep the animals locked in at night and Gypsy is always very keen to get outside in the morning. She goes to the door anywhere from 4–5 am and meows to be let out, while we lie there saying, 'Yes, when we get up we'll let you out' and remain snoozing in bed.

Kooch is a little uneasy around Gypsy the cat

However, clever little Gypsy quickly figured out that if she goes up to Kooch and annoys her, Kooch will whine and whinge a little, then get up out of her bed and go to the back door. As she's incontinent, we always spring out of bed and open the door for her in case she has a little accident, thus letting Gypsy out as well.

Jo Shirlock
Morphettville, Adelaide
South Australia

Not a drop to drink

It was a Saturday evening and we had friends over. After feeding our two dogs Fritha and Tuffy their dinner, we sat down for ours.

We had been at the table for five or ten minutes when I felt a tap on my leg. Imagine my surprise – and everyone's amusement – when I looked down and there was Fritha sitting beside my chair, gazing up at me with an empty bowl in her mouth.

I knew by the look on Fritha's cheeky little face that I was in disgrace. I always keep the dogs' water bowl full and she was saying, *Haven't you forgotten something, Mum?* The oversight was soon remedied!

Faye North
Katikati
New Zealand

Chub-Chub's personal servant

A few years ago when I used to raise mice, I came across a hugely obese mouse for sale in a pet shop. It was love at first sight. I am good friends with the pet shop staff and they gave her to me for free, so I left the shop holding the newly named Chub-Chub.

At that time I had a one-year-old mouse called Splodge who I decided would make a perfect friend for Chub-Chub. When I brought Chub-Chub home she was already quite old and found it very hard to move around. I was worried about her, as every day she looked sicker and kept losing her balance and ending belly up on the floor of the cage.

Then one day when I came to see her she waddled furiously over to my hand to be picked up. She had never been that energetic before, so I decided to watch her to find out her secret. I managed to catch

Splodge in the act of bringing her food from the bowl and rolling her over whenever she slipped. I didn't know mice were that smart! They both lived for ages after that and I still giggle to myself when I think of them.

Siobhan Greening
Cordeaux Heights, New South Wales
Australia

The great escape

We have a flat-coated retriever called Newton and a chocolate Lab called Sumo who kept escaping from our garden, but we didn't know how.

Then one morning we found them in the middle of an escape. Newton was digging a hole under the fence. Every so often he would stand back, and then Sumo – who at the time was chubbier than Newton – would stick his head into the hole to see if he could fit through. If he didn't fit, he would turn to Newton and back away from the hole, and Newton would dig some more. So we found out the secret of our dogs' great escape!

Emma Bridge
Horseheath, Cambridgeshire
England

Write to me ...
Emma Bridge
Bridge Corner, West Wickham Road
Horseheath, Cambridgeshire
CB1 6QA
United Kingdom
or email Emma
bridgeemma@yahoo.co.uk
or dumb_non_blonde@hotmail.com

Escape artists: Newton and Sumo when he was a puppy

7

Smart animals get revenge and outwit others

A clever old dog

At any one time, we always have two or three dogs of varying ages. By the time our ridgeback cross Rhodie became the aged one, she had lots of tricks up her sleeve.

Rhodie was very particular about which chair she liked to sit in, her choice depending on its closeness to the fire. Humans were very easy to move: a mournful look and lots of crying always did the trick.

Dogs, however, required a different tactic. She would amble to the front door and then begin to bark frantically. As the door was opened and the other dogs rushed out to investigate, she would calmly turn round and take up residence in the previously occupied chair.

Jill Grist
Augusta
Western Australia

Grab that cat!

Our orange house cat is called Sonny, so named because he was the first boy in a family of four girls. After a few years of being the only pet, aloof Sonny was introduced to a new family member, a black and white springer mix named Missy. Despite Missy's eagerness to be pals, Sonny gave her the cold shoulder and pretended she wasn't there.

Sonny is so much like the overweight cartoon cat Garfield that he could play the part. He has lost some of his roundness since this incident, but he was pretty round at the time.

Missy has the exact opposite personality of Sonny; she is bouncy, outgoing and always alert. She is like another busy 'person' in the home, responding to everyone's comings and goings and every noise, and generally keeping tabs on everything. Sonny, on the other hand, has no desire to move except when food is involved. But every now and then, when the door is open and everyone is distracted, he makes a symbolic attempt to escape.

One day shortly after Missy's arrival, I was holding the door open as I called to the girls playing in the yard. Suddenly, an orange blur went past my legs. Realising it was Sonny heading down the steps of the deck, I yelled, 'Grab the cat!' Seconds later I was nearly knocked off my feet by a black and white blur pushing past me and out the door.

Sonny had only gone about six feet on the grass when Missy's paw came down on his neck. He flopped over on his side and lay there, making no attempt to run, while Missy looked over to where I was standing by the door as if asking me, *This cat?* or *Now what?* Sonny, obviously thoroughly disgusted, seemed to be thinking, *Whose idea was it to get a dog?*

The girls and my husband were watching from the yard and Grandma, Grandpa and I from the house. We all broke into laughter; it had happened so fast and was so funny.

Since then it's happened again a couple of times. It's a great game for all – and just as funny every time!

Kari Jones
Sturgeon Falls, Ontario
Canada

Clever girl

Every once in a while you meet a dog with character who really stands out. When I was a child our family had three dogs: Ajax, an Irish setter; Bella, a dachshund; and Pasha, a miniature dachshund. Now, it was always obvious that Bella was the smartest by far, and she demonstrated this every day.

Each evening when all three dogs were fed, Bella used to wolf down her food while the two males were still working out where to begin eating. Then, bored and probably envious that she had finished her food while the others still had theirs, she would walk down to the gate and begin barking as if someone was coming down the street. The other two dogs would leave their food and go down too, barking. While they were still trying to work out what was going on, Bella would run back to the food bowls and quickly finish off what was left.

She used this trick every day but the other two never caught on.

She was well and truly the leader of the pack. When it came to tracking games, she could follow any trail out in front while the

Coco made his feelings clear

others dogs just followed her. She was the one who decided who was friend or foe. She knew how to stow away in a car so that we didn't realise she had come out with us until it was too late. She was definitely the boss and quite fearless.

Bella died at 15 years old with the usual dachshund problem – her back – but is still remembered decades later as an important part of our family history.

Marie Belcredi
Epping, New South Wales
Australia

That's MY toy!

Coco the cat enjoyed being our only pet. He was a playful kitten who loved toys. When we gave him a pre-loved mouse-shaped cat toy that we found in the garden of our new house, he took pleasure in batting it round the dining room, or holding it with his mouth and front paws while his back paws scratched at it furiously.

As he grew older, Coco became more serious. The cat toys were abandoned for more adult pursuits such as stalking real-life mice or birds.

Coco became more serious still when he stopped being an only cat. The arrival of Dusty the kitten when he was three years old pleased the rest of our family but not Coco. He showed us in various ways that he did not appreciate this interloper, such as raising his hackles and hissing at her.

Dusty was every bit as playful as Coco had been, so naturally we gave her some cat toys, including Coco's old toy mouse. It seemed that the sight of Dusty playing with his long-abandoned toy was too much for jealous Coco. When Dusty had temporarily discarded it,

he picked up the toy in his mouth and trotted purposefully outside. Once there, cunning Coco sneaked under the veranda and hid the toy in the dense shrubbery.

It was the last time Dusty played with that particular toy, for it never saw the light of day again.

Linda Richardson
Wellington
New Zealand

Gentle Bluey

The most intelligent cat I ever had was a big gentle black and white fellow we called Bluey.

When I first set eyes on this lovely cat I thought it was the cat next door as they were identical. Some months later my neighbour said, 'Your black and white cat is often in here stealing Charlie's food.'

'I haven't got a black and white cat!' I answered.

With that, we looked into the garden and for the first time I saw the two cats together, sitting in the sun like identical twins.

Clearly, someone had loved and cared for Bluey, as his pink nose had been tattooed blue to prevent skin cancer and he was desexed and in very good condition. I did the usual advertising, but no owner was found so I decided to take him on.

I had two other cats at that time and both of them were a bit miffed about me taking on a stray. There were lots of challenging looks and much posturing, but Blue was way too smart for them. When faced with one of their challenges he would adopt a humble demeanour and avoid eye contact. By simply sitting quietly and looking at his feet, he seemed to offer no threat to the challengers and they let him be.

Lovely Bluey was a special cat

Blue was much bigger than either of my other cats and could have shredded them if that had been his way, but he was too intelligent to get into a confrontation. He quickly worked out who sat where and who owned which pillow, and found his own little piece of comfort where he wasn't impinging on space already spoken for.

Blue was one of those cats who seemed to know what you were saying. One evening as he was about to enter the lounge through the window, I noticed one of the other cats lying in ambush behind a pot plant. I called out 'Blue!' to get his attention and he looked at me from the windowsill. I said, 'Ambush' and pointed to the villain behind the pot plant. Blue's intelligent eyes immediately sought out the perpetrator and he looked back at me with a knowing smile. He then went round to the back of the house and came in through the cat hatch – leaving the villain totally perplexed when Blue walked into the lounge from the other direction!

I've had many cats over the years – usually two or three at the same time – and all were lovely, but Blue had something special. He'd look into your eyes and you knew he was a soulmate.

Last year I was interstate for about a month during my father's passing. During that month my neighbour fed the cats as usual. I returned home in the early evening and was met by the three cats, who were all very pleased to see me. I cuddled them each in turn and then fed them. But Blue didn't touch his food; instead he went outside and meowed, behaviour I had never seen before.

I went out into the fading light and picked him up. He was as light as a feather. We had a little chat and I put him back down. He started to walk across the garden, and when he was about ten feet away he paused and looked back at me. We held eye contact for quite a while and then he went on his way. I knew I'd never see him again.

I didn't know his age when he arrived and we had him for about ten years, so he may have been 15 or 16. He'd been looking old during that past year. I think that Blue had waited a month until I returned; he knew he was dying and just wanted to say goodbye.

I thought it strange that this special cat had gone off to die within a week of my father's death. Dad adored cats and I like to think of them together, two intelligent beings happy in each other's company.

Terri Cracknell
Manly, New South Wales
Australia

Write to me ...
Email Terri
terazzo@iprimus.com.au

Henny Penny and the bear

At 20, Henny Penny was showing her age. Her bantam feathers were a little grey and in places a bit thinned out but, although her eyes weren't quite as bright and her eggs were laid less frequently, I was very fond of her. She'd been a terrific mother.

I remembered one occasion when I'd watched her from the cabin door. The rain had soaked the grass overnight and she'd picked her way through the longer stems with clear distaste, lifting first one foot and shaking the water off, then the other. Her clutch of young chicks scurried along behind her across the wet lawn.

Henny Penny knew that her chicks were getting just as wet as she was, and just as cold. Despite their very loud protests, she herded them into a manageable group and plonked herself down on top of them. She was going to warm them up. I sighed. That had been a few years ago, and her last set of chicks.

Nowadays, she still clucked and pecked about as she always had, but sometimes she seemed to peer wistfully behind her, as if expecting to see another group of youngsters who needed her.

Frankly, I didn't particularly want any more chicks. I had enough hens, and the rooster I'd recently purchased was one too many – or so it seemed, as he often liked to crow well before sunrise. Besides, I thought Henny Penny was too old to raise another brood. I didn't know what age a hen had to be before she could no longer lay fertile eggs but, just in case, I shut the rooster up in a pen of his own.

I lived on Kootenay Lake, just a hundred yards up off the water at a spot called Six-Mile. Behind the cabin, a meadow stretched high towards the trees and brush covering the mountain. My garden sat to one side of the house, with the hen house just beyond that, close to the trees so their shade would keep it cool in the summer.

One day that autumn a bear turned up. Every day it hid in the bush, just out of sight. Normally, I wouldn't have worried. Autumn is the time when bears eat everything they can. They look for berries or ripe plums and apples, rich food to quickly fatten themselves up for the coming winter's sleep. On the slopes of the mountains beside Kootenay Lake there were lots of berries and bugs to feed them, particularly huckleberries higher up. Most of the homes along the lake had at least one fruit tree in the yard, which occasionally drew the attention of a passing bear. However, once it had eaten its fill it usually moved on in search of more food, so I was a little concerned when this one stayed. Each day it lurked beneath the trees, watching the hens. This bear wanted meat. Chicken meat.

The next day Henny Penny was gone, along with the rooster. Only scattered feathers lay outside the chicken shed, the wire fence for the rooster's pen crushed to the ground. I immediately phoned a game warden to have the bear removed, cursing myself for waiting too long. A bear that stalked domestic animals would eventually go after a human. Besides, it had taken Henny Penny, my favourite. The bear had to go.

Two officers trapped the bear the same day, and waved as they carted it off. 'We'll take him up the Ymir logging road,' they told me. 'No one for miles, and lots of huckleberries out there.' The animal – a young one – glared out at me from the back of its cage as the truck drove off.

Good riddance, I thought, thinking of Henny Penny. She'd had a good life, I told myself. But somehow the meadow and the garden seemed forlorn without her familiar presence.

Two weeks later, Henny Penny trooped out of the brush, a clutch of chicks at her heels. She preened and strutted, sure that I'd be pleased to see her. She paraded her new youngsters before me, knowing I'd admire them. I clapped my hands, delighted. She'd gone to ground, hiding until the bear was gone, then laid her eggs and hatched them, returning only when she was sure it was safe.

She had one more set of chicks after that, courtesy of the rooster – who I never did see again, though I often heard him off in the trees somewhere, crowing to the mountain about his freedom. Sadly, a year later I woke up one morning to find Henny Penny unmoving beneath the raspberry canes, one of her favourite spots. She was a very smart old bird.

Sharman Horwood
Seodaemun-gu, Seoul
South Korea

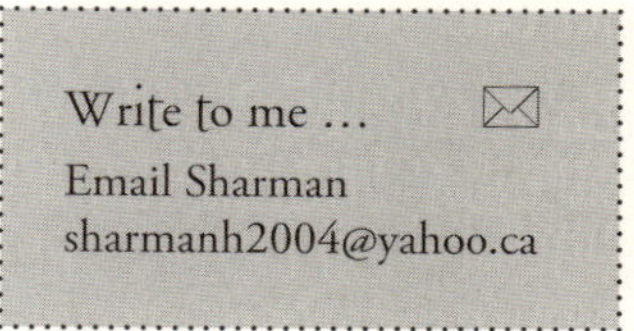

Sharman Horwood is a Canadian writer living in Seoul, South Korea.

Inseparable friends: Chaos and Duke

Duke's sneaky treat

Duke was a beautiful black and tan one-year-old German shepherd cross who had become more than just a pet, he was a treasured family member. So when my sister Renee came home one afternoon and proudly announced that she was moving out and taking Duke with her, the family was devastated.

A few weeks after Renee moved out, the house still felt too empty. Something had to be done. A couple of weeks later we adopted Chaos, a two-month-old black Lab/German shepherd cross.

Chaos's name seemed very appropriate, because he was a little devil! He would eat anything in his path including plastic, tin cans and wood. He never shared and always got his own way. At first we were afraid that Duke would be jealous of the new puppy, but he took to Chaos like a father and soon they were inseparable. Renee didn't live far away so twice a week we had a 'doggy sleepover'.

On one doggy sleepover night the dogs were getting settled after doing their business and getting their treats, when I noticed that Duke had a lump on his mouth. I thought, 'Oh my goodness, how will I break the news to the family that Duke is sick? How long will he have to live?' I felt his mouth and to my surprise it wasn't a lump – it was a bone that he was hiding from Chaos!

Duke moved closer to my pillow, then stretched out and pretended to fall asleep, with the bone still in his mouth. Of course Chaos fell asleep very quickly, being a puppy. As soon as Duke was sure that Chaos was asleep he opened his eyes and started chewing on his bone in peace. I couldn't believe it! How smart was that?

Duke is currently 35 pounds and wouldn't be able to get away with that trick now because Chaos is not only getting smarter as he gets older, he now weighs 65 pounds and still doesn't like to share! We are currently working on that as we've just adopted a kitten – Katara – and it would be great if Chaos learned to share with her.

Rhonda Roebotham
St John's, Newfoundland
Canada

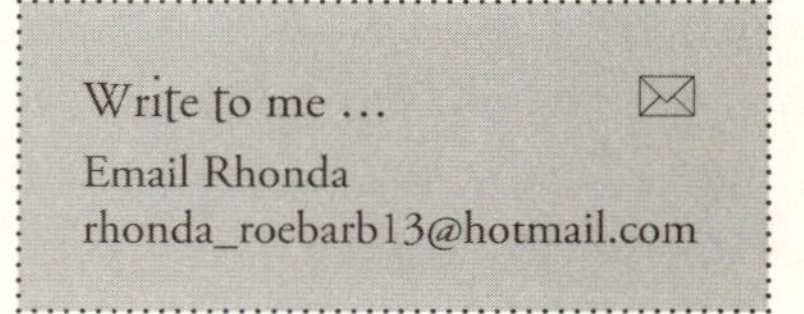

A feline telltale

We had three cats, one for each of our three children. Josie was a long-haired dark tortoiseshell female, Pink was a large short-haired black tom, and Ricky their mother was a dark tabby.

Because I worked full time and the children attended school, I had a set routine to make sure the cats were fed and all outside before I left for work. Three dishes of food were placed in front of the kitchen window. While the cats ate, I checked that all other

windows and outside doors were shut. Going back into the kitchen and seeing that the cats had finished their food, I'd push the window open wide and clap my hands twice, and immediately the cats would jump out the window one after the other. It worked a treat; the cats knew exactly what was expected of them.

One morning when I returned to the kitchen after checking the windows, there were only two cats waiting to go outside. I didn't worry, as I thought Ricky must have already jumped out the window. However, when I came home I was surprised to find her stretched out on the carpet, soaking up the afternoon sun in the lounge. I decided I would have to be more vigilant tomorrow.

The next morning I followed the same routine, but Josie was the only cat waiting by the window to be let out. I decided to find the other two.

When I went into the second bedroom, I saw Pink sitting beside one of the beds. 'What do you think you're doing there?' I asked him. Pink looked straight at me, lowered his head slightly, looked under the bed and looked straight at me again. I couldn't believe it – was he telling me that Ricky was under the bed? The message seemed clear enough, but I had never had a cat communicate with me in this way before.

Sure enough, when I lifted the bedspread there was Ricky, crouched down and trying to hide.

Pink hadn't been trying to hide from me when he knew I was looking for them. He was simply waiting to show me where Ricky was hiding, knowing they always had to go outside. Pink was 'telling tales' on Ricky!

Faye Ross
Ngatea, Hauraki Plains
New Zealand

Dogs versus cats

I have a bearded collie cross called Molly who is very picky about what food she eats. She took a liking to the next door neighbour's cat's food. Molly would often run through the hedge separating the two houses to get through the back door and eat the cat's food.

Soon the cat caught on and decided to stand next to the back door to protect its food. Unfortunately, the cat wasn't too bright and so, while it was guarding the food, Molly ran out our front door, along the road and in the next door neighbour's front door, and smartly ate the food behind the poor cat's back.

Lindsay Furze
Tollerton, Nottingham
England

Sneaky Tiger is foiled

I had come back from walking Bonnie, my dog (and the bane of my cat Tiger's life). Without realising I hadn't shut the front door properly, I went and made a cup of tea.

The cats are *absolutely not ever* allowed outside the front of my house because the road can be very busy, so of course Tiger has made it his life's ambition to see what is out there.

I was drinking my tea when, out of the blue, Bonnie came running into the kitchen and barked at me. Then she started whining and tugging on my trousers with her teeth. As this was not her normal behaviour I went to see what was wrong.

I got into the hall just in time to see Tiger sneaking out of the front door, which was slightly ajar. Bonnie went after him and tried to herd him back into the house with her nose, whining all the while.

Tiger seems to be wondering, 'Is anyone watching?'

To say Tiger was annoyed that he had been caught out was an understatement. He gave Bonnie a look of pure loathing, slapped her on the head and went off upstairs in a right huff. I, on the other hand, gave Bonnie a big cuddle and told her what a good girl she was.

My dog is not nearly as dumb as she makes out.

Helen Miles
Cardiff
Wales

Write to me ...
Email Helen
leopardusweidii@yahoo.co.uk

8

Smart animals don't go hungry

Great teamwork

My son Brad was doing some carpentry as part of a home remodelling job. The couple living there had told him that squirrels kept getting into their bird feeder, despite the fact they had done everything possible to prevent them from doing so. They had never actually seen any squirrels at the feeder and had no idea how they got to it.

The owners were away at work and Brad was sitting at their breakfast nook table having his lunch, when he saw four squirrels climb out on a branch hanging above the bird feeder. Their weight caused the branch to sag enough for the first squirrel to jump onto the feeder. After a feed he jumped down, then climbed back up the tree and onto the branch to be fourth in line, so that the next squirrel could jump onto the feeder. This continued until all four squirrels had had a meal.

Chuck Walker
Heyfield, Victoria
Australia

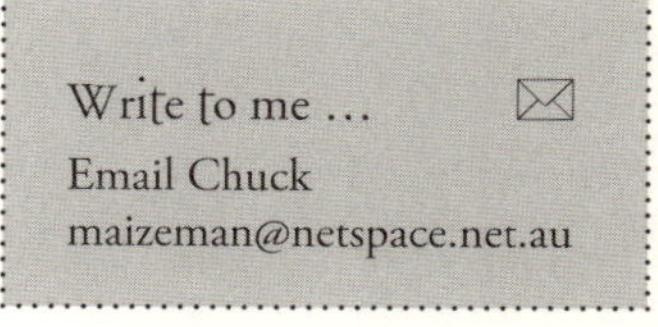

Stop, drop and roll

I once had a wee grey pony named Soreya who loved to eat the juiciest clover grass – so much so, in fact, that we had to fence her away from it so she wouldn't get sick. If a pony eats too much, its feet swell and this may result in founder (also called laminitis), a crippling and painful disease.

I would let Soreya eat in the big paddock for an hour or so a day among the tastiest grass, then I would move her and leave her fenced in the small paddock overnight. However, every morning I would find her out in the big paddock. This intrigued me, as I knew she couldn't jump high enough to scale the fence, not with her belly hanging so low!

One night I decided to watch, and stationed myself outside where she couldn't see me. The porch light was on, giving me just enough light to see into the paddock.

The next thing I saw was the cheeky wee thing strolling up to the fence, sinking down onto her knees and starting to roll. She rolled right under the fence – which consisted of two rails that gave her just enough room to manoeuvre under – and stood up on the other side, triumphant.

Straight away it was head down, bum up, and she began to eat into the night. Needless to say, we added more rails to the fence – much to her disgust.

Bess
North Canterbury
New Zealand

Write to me ...
Email Bess
gingergirl_13@hotmail.com

One for you, one for me

One afternoon my wife Pat made some little cakes. Our Jack Russell terrier Susie, unaware that we could see her via a large mirror, jumped on a chair, stole a cake, jumped down and deposited it in front of the sleeping cat, then returned and stole a second one, took it to her basket and ate it.

We didn't have the heart to chastise her: in light of such generous skulduggery we thought she deserved to get away with it.

L L Dunn
Gosnells
Western Australia

The cat kibble conundrum

Maxwell is a black cat who is highly motivated by his stomach. We didn't realise just how motivated until we noticed we were having cat kibble (dry cat food) storage problems.

After our two cats' last annual check-up, our veterinarian advised us to limit their meals to twice a day and to use 'portion control'. As a result, one or both of them were gnawing their way through the bag of kibble in a vain attempt to convince us that they were in fact starving.

I upgraded the kibble storage to a file-folder-sized plastic Rubbermaid box topped by a removable lid with two latch handles, one at either end. To secure the lid you snap both latches down, and to lift the lid you snap the latches up.

The next two mornings when I went to the box to feed Maxwell and his brother Jake breakfast, I noticed that the lid was skewed

sideways. I assumed I had not snapped the latches down, and as I was reminding myself 'Remember to snap the latches shut', I thought how effective this Rubbermaid box was. It kept the kibble fresh, it stacked nicely, it provided easy access to the kibble and it was a neat way to prevent the boys from eating themselves into absolute butterballs.

That evening, I heard an odd rustling coming from the bottom of the stairs in the boys' dining area. I peered down the stairwell to witness Maxwell standing on his hind legs, nonchalantly flipping up the second latch on my new storage box with his front paw to match the other latch he had already flipped up. Bracing both front paws on the edge of the lid, he pushed the lid sideways and helped himself to more dinner. He glanced up, saw me observing his commando tactics and gave me a sheepish meow.

I solved this problem by putting a box on top of the Rubbermaid lid. The next morning I found the lid skewed sideways once more. Not only is Maxwell a highly motivated cat, he is a strong one. Finally, I placed a whole pallet of canned tomatoes on top of the lid. For the next few days, I would see him, standing and looking at the pallet of cans, his tail twitching as he contemplated this new conundrum. He hasn't figured this one out – yet.

May Yeung
Calgary, Alberta
Canada

Clever Maxwell

Midnight feast

Billie the beagle was the sharpest thing on four legs. She never missed a beat and was always the first animal on hand at any sign or smell of food.

One night we all went out for a meal, leaving Billie in her usual resting place on her beanbag under the coffee table in the kitchen. Two hours later we returned home. Billie lay just as we had left her, snoozing on her beanbag, the epitome of drowsy innocence.

Someone suggested a cup of tea. I opened the fridge to retrieve the milk, only to be confronted by an almost empty refrigerator. Almost everything in it that was edible had gone! We were all flabbergasted and puzzled. No one but Billie had been in the kitchen. How could a dog open a refrigerator?

Days passed and each day something else would disappear. Then one sunny morning, all the doors wide open, I walked into the kitchen just in time to hear shouting from down in the yard. I rushed through and out the back door, to be greeted by my husband's voice calling, 'Come here, you little …!' – just in time to see Billie disappearing round the back of the house with a whole dog roll in her mouth, running like the wind.

So we set up a stake-out. Sure enough, the moment the kitchen was deserted, madam proceeded to the fridge. With her nose deftly poked under the bottom of the door, she flicked it open. Simple. And caught in the act.

From that time until she died aged 16 years we had to put a 'stretchy' band on the handle and hook it back to the wall, so ending the midnight feasts of Billie the beagle!

Barbara Molloy
Christchurch
New Zealand

One-Eye

I'm not very good at milking – opportunities had been rare in the city – so, when something nudged my foot, I assumed it was the cow objecting. But it wasn't. Bette (she had Bette Davis eyes) was placidly ignoring my unskilled squeezes. She had a very tolerant nature. She chewed on a bit of hay, eyes rolling every so often, which I'm sure was her way of wincing when I pulled a little too enthusiastically.

New Zealand's Tararua Ranges glittered in the distance, their peaks dusted with winter snow. I was visiting my friend Lyn, and enjoying the clean air, new lambs and lush greenness of the countryside. Lyn had chuckled when I offered to milk the house cow. She knew just how much – or, rather, how little – milking I'd done in my life. But she was willing to let me try.

I was determined to master this skill, and persevered. Something nudged my foot again, this time a little more strongly. Bette's hoof was nowhere near. I looked down. There, at my feet, a one-eyed hedgehog leered up at me. In Canada we don't have hedgehogs in the wild. I'd never seen one before. Her face was a little skewed; whatever had damaged her eye had affected her mouth too, so the lip on one side was lifted, like she was snarling. Or trying to grin pleasantly up at me.

'What do you want, girl?' I asked her. She wasn't afraid of me, and she certainly hadn't bitten my foot. She was looking for something. And as the barn cat curled around the corner I realised what it was. Milk. That glorious drink most animals love. Lyn kept a flat pan near the milk bucket and usually squirted some into it for the cat. Apparently, One-Eye expected her share, too. I laughed and obliged, quite delighted with the imperious little soul patiently waiting for this stranger to learn exactly what her farm duties entailed.

She drank what I put in the pan, and then waddled off. For the next year, Lyn told me in her letters, One-Eye came out every day for her bit of milk, and then scuttled back under the barn. It was warm under there, even in winter, and a safe place for her to make her home.

About a year later, One-Eye died. After Lyn found the small hedgehog's body she heard a noise, like the keening cry of a small animal. Not a kitten – those were warm and protected under the porch. This was something different, and it certainly didn't sound like a rat. She dug around in the back of the hay bales. There she found a baby hedgehog, One-Eye's kit. Before she'd passed away, the hedgehog had brought her baby out from under the barn and tucked her into a place where a milk-providing human would find her. It was cheeky. But smart.

Lyn obliged. She took the baby to the nearest vet and he found a good home for her.

Sharman Horwood
Seodaemun-gu, Seoul
South Korea

Write to me ...
Email Sharman
sharmanh2004@yahoo.ca

Sharman Horwood is a Canadian writer living in Seoul, South Korea.

Thomas's summer treat

It's summertime again – a time of long hot days to laze by the river or on the beach, a time to enjoy salads, ice creams and long cool drinks. On days like this, my husband André enjoys a tall glass of iced coffee.

Thomas, my nine-year-old Lowchen, normally takes very little notice of any activity in my kitchen until it comes to his mealtimes. He certainly makes no fuss when I boil the kettle to prepare a hot cup of tea or coffee.

However, one summer day a few years ago, after I given André his iced coffee, Thomas suddenly got up from where he had been sleeping and went over to sit beside my husband, looking at him pleadingly. Before long those big brown eyes had worked their magic and André offered him a teaspoonful of his coffee. Thomas lapped it up and waited for another, and my husband obliged. After two spoonfuls Thomas was satisfied and went off to resume his afternoon sleep. This was repeated every day when my husband drank his iced coffee.

As the days shortened and cooled down, André returned to drinking his ordinary cup of hot coffee. Thomas never attempted to move from his place or ask for a taste.

On the first hot day of the following summer, I asked André, 'Would you like a glass of iced coffee?' As I took the long glass from the cupboard we noticed that Thomas was again sitting by André's chair waiting for his small summer treat.

This happens every summer and we find it quite amazing that Thomas has such a good memory!

Mary Delcourt
Ashby
Western Australia

Dinner's late today

One day I dozed off in the lounge after a relaxing cup of coffee. I awoke what I thought was a short time later to a persistent tapping on the French doors.

I looked out towards the noise, and was confronted by an array of concerned-looking geese and turkey faces peering through the French doors from the deck. All the geese had their heads tilted to

Where's our dinner?

the side and the turkeys had their beaks against the glass; they were not uttering a sound. I was so taken aback by this one-off vision that I forgot to grab the camera.

I think there is only one explanation: they had come looking for their dinner, which was two hours late. I like to think they were also a bit concerned about me, not just their food!

As soon as I showed myself and began talking to them they all started honking and gobbling, making a wonderful noise. They never cease to amaze me.

June Spragg
Warkworth
New Zealand

Write to me ...

June Spragg
453 West Coast Road
Ahuroa
R D 1
Warkworth 1241
New Zealand

Second helpings

After noticing that our cat had been wandering the streets a lot, we decided to get him neutered.

A couple of days later, a friend of ours came across his neighbour with a distressed look on her face. 'I think someone has neutered our cat!' she exclaimed.

Putting two and two together, our friend soon realised that the 'stray' cat Mortamar whom his neighbour had been looking after was actually our family pet Pickles. For the last four months he had been eating at our house, then wandering down the road for a second helping.

No wonder he was putting on weight!

Jess Yarrall
Blenheim
New Zealand

The good stuff

This is the story of Vanna, a pure-bred Samoyed husky (white and biscuit), who won best in her class with the Canadian Kennel Club in 1995.

I was married at the time of this story to a former East German who had defected to Newfoundland. Over the years, he was always saying things like, 'In wartime you always eat the good things first 'cause you never know when you'll get them again.' Anyway, Vanna apparently listened.

Vanna was the type of dog who always barked when someone came to the door and she was always there to help you open it. One day my husband was getting himself a snack in the middle of

Vanna was a good listener

the afternoon, which consisted of German bread smothered in real butter and then some German sausage, salami, black forest ham, etc on top.

The doorbell rang and Vanna, who was lying on the living room floor, did not get up to answer it. My husband went to the door, and was there a few minutes as it was a courier delivering a package. By the time he came back into the living room Vanna had managed to eat all the meat off about six slices of bread, and had all the bread on the floor and was starting to lick off the butter – because in wartime you never know how long you have to eat the good stuff!

She was such a good listener and a wonderful dog. I was very lucky to have had her in my life, even if only for a short time.

Patricia Collins
St John's, Newfoundland
Canada

Do you have a burning question?

Have the actions of an animal baffled you recently? Chances are someone else has encountered the same situation. Send us your question and we may publish it in a future edition in the SMARTER than JACK series.

Readers will be invited to offer solutions and maybe your question will be answered. You'll also receive a complimentary copy of the book in which your question is published.

For submission information please go to page 146.

9

Smart animals take control

Not without me!

The Staffordshire bull terrier arrived in our street and happily divided his time between our neighbours and us. We knew his time with us was limited and that his owners would claim him, but he was a generous little chap and we enjoyed having him around.

When my husband Phillip arrived home from work he changed into his jogging clothes and set off along the bush tracks behind our house. The Staffy joined him and ran the ten kilometres easily, leaping over bushes and enjoying his run.

The following day was Saturday, and my husband went running early in the morning. He looked for the dog but it wasn't there, so he went alone. When he returned, the Staffy was waiting for him and scolded him severely for going without him.

When Phillip rose early the next morning, there was the Staffy asleep on our patio with Phillip's jogging shoes held firmly between his paws.

Jan Schelle
Clifton Beach, Queensland
Australia

Write to me …
Email Jan
j_pschelle@hotmail.com

Alex won

I always knew there was something unique about my 24 pound tabby cat Alex. Just looking into his big yellow eyes you can tell he knows something you don't.

I didn't realise just how intelligent he was, though, until about three months ago when my bedroom door was mysteriously opened after I'd shut it – this happened numerous times. I thought maybe it was the wind, or the door just wasn't good at staying shut. But then about two weeks after this began happening I discovered the real problem.

I got ready for bed about 10 pm. Alex had been following me around all evening and, as he usually keeps me up during the night, I decided to shut the bedroom door before he tried to sneak in. The door was shut good – real good, I made sure of it.

As I was trying to go to sleep I kept hearing scratching noises at the door. Of course, I knew it was Alex scratching to get in so I ignored it. About 30 minutes later I heard the doorknob turn. I got up to see who it was, and there at the door was Alex, waiting to come in. I looked around to see if anyone else was awake, but to my surprise the lights were off and there was not a peep in the house.

Alex had been the one opening the door all along, just trying to get in the room to me. He stays there now. I figure I might as well let him sleep with me – it saves me having to get up to shut the door behind him!

Audrey Abrams
Edmond, Oklahoma
United States of America

Darling Webster charmed everyone he met

Webster worked his magic

Shock swept Liz's face. 'Oh, you poor darling,' she breathed, eyes resting on the twisted paw placed firmly on her knee. Fixed by Webster's beseeching gaze, she hardly seemed to notice her biscuit slipping soundlessly towards his waiting jaws.

Webster, a displaced black and white terrier cross who'd limped his way into my life early one cold April morning, had found an unfailing way to use his past misfortunes to his advantage. Visitors, he knew, would melt at the sight of the grossly deformed leg which, sometime before, had been smashed by his abuser and left without

treatment. Drawing attention to it was a fail-safe way to receive generous titbits, whispered sympathies and copious caresses.

It had not taken long to work out what had happened to him. The vet confirmed that the leg had been broken and left untreated, and – there and then – Webster worked his magic: the vet's parting words were, 'This one's on me.'

I soon discovered that Webster (who at this stage was not named because my husband declared that he must be rehomed) was terrified of men. Even when he began to understand that most men were safe, he would suddenly disappear into hiding. The white tip of his tail was usually the giveaway, and I would find him under a bush or behind a chair, trembling in anticipation of the next beating.

As hiding stopped, the nightmares began. Alerted by a series of single barks, I would run down and sit with him until his lingering dreams subsided. But, as the terrors began to pass, Webster was learning the power of his deformed leg.

Perhaps it was because I tended to tell his story to any dog walker or visitor who asked why he was limping. Soon he learned to anticipate these conversations, and would present his leg unprompted. Webster had discovered his audience and knew just how to play to it. Even with us, he began to use that leg to his advantage.

Eighteen months on and feeling more secure, Webster began supplementing his walks with a series of carefully planned escapes. He would choose a hidden part of the fence and begin work on a hole, gnawing through the wood if he couldn't tunnel under it. Then he would disappear. After hours of fruitless searching we would see him at last, jaunty and triumphant, tail held high, making his way home. But the moment he spotted us his countenance would change. The waving tail dipped, his head dropped and that characteristic limp would suddenly get much worse. Who could be cross with him?

Webster – he was just irresistible. A terrier through and through, he was a true 'earth' dog, tossing clods and divots, preferring to scrape at the soil rather than lie on the lawn, and frequently disappearing down foxholes. He bared his teeth at the flies on his bone, the cat on my lap, and all birds bigger than magpies because he was once attacked and dunked by a territorial swan. He disdained toys for knotted rags and growly rough-and-tumble terrier games – except at the boarding kennels where, inexplicably, he retrieved a ball just like a proper dog.

Most memorable is the love, devotion and gratitude he lavished on me. Webster did not offer obedience easily; at the kennels, he would sit down when called from the exercise field, and refuse to move until clipped to a collar and lead. ('But you can't be cross with him,' the proprietor informed me, 'it's Webster.') Nonetheless, he would do for me whatever I asked of him, sometimes taking a moment first as if to show that he was obeying by choice. He was my constant companion, bestowing a love and empathy that went way beyond the opportunism so often reserved for guests.

Towards the end of Webster's life, we had a patio built. He watched intently, brick by brick, and promptly sat upon it as the last slab was laid. He had no doubt, even for a moment, that it was purely for his benefit. The last photographs we took of him were on that patio during a barbeque. Needless to say, he spent much of that evening presenting the leg to our hapless guests, who promptly rewarded him with huge chunks of chicken, sausage and best rib-eye steak.

Shortly after that evening, Webster became very ill. He simply wished to be still, and was able to lick my hand but uncharacteristically refused food and drink. He seemed to understand that his time had come – even though I did not – and took on a whole new serenity and dignity.

When it was all over and Webster lay silent and still on the vet's table, the last part of him caressed was that old twisted leg. Perhaps I was thinking I would never see the like again, at least in this life. I was wrong. I see Webster often. Most nights, at some stage, he limps his way through my dreams, and then it's just as if he has never gone away at all.

Irena Szirtes
West Smethwick, West Midlands
England

Write to me ...
Email Irena
b.glad@btinternet.com

Person-tamer

In January 2005 I had a visitor. I walked out on my farmlet one morning to feed the poultry and found I was being implored for aid by a young magpie. I gathered his parents had decided he was big enough to take care of himself and had left him to make it on his own. As far as he was concerned, food was scarce and life was hard, and he'd often seen me ambling past the trees in which his nest was situated.

His parents must have told him he didn't have to worry as I wasn't a predator. So Junior had gone a step further. If I wasn't a predator, could I be a provider instead? He hopped hopefully up and begged. I went inside, and returned with half a slice of bread.

I bent and extended the bread towards him, and he bravely hopped right up, took it gently from my fingers and hopped off to enjoy his windfall. After that he appeared every morning, and some evenings, for a month. I gathered that I was a supplementary supply. I'd give him the energy to get started in the morning, and a bit extra in the evening if it hadn't been a good day.

I initially had no idea how long he'd do this, but I've always liked the white-backed magpies here in Hawkes Bay and had no problems with helping out. After four weeks Junior had grown more accustomed to finding his own food, because he stopped begging every day.

Then in March I impressed our meter reader when he was standing just outside the back door talking to me. From behind him came an imperative warble and he turned to discover a magpie glaring at us. Where was his bread? I excused myself, got bread and proffered it. To the meter reader's astonishment the magpie hopped right up to us, grabbed the bread and left.

Was that a magpie I'd tamed? he asked. I suspect it's more that I'm a human the magpie has tamed. As I discovered later, the magpie had extended human taming further than anyone had realised. Friends from the house bus on the corner of my farmlet saw me feeding my friend and commented that they'd been feeding him too. Then I happened to be talking with next door, and found he'd arrived on their doorstep and they were feeding him as well.

So, while I think he started here first, he worked out quickly that owning humans freed a bird to do all kinds of things besides looking for food. What I'm considering is him teaching all his offspring to expect this too. Just what we need in our village – a flock of white-backed magpies who've tamed the residents to feed them on command!

Lyn McConchie
Norsewood
New Zealand

Stay home!

We had a ten-year-old miniature fox terrier named Digger. We loved him so much and hated leaving him when we went travelling. He used to get very sad and upset whenever we got the suitcases out because he knew that we were leaving him behind.

One Christmas we were getting ready to go on holiday. The suitcases had been packed the night before so that we were ready to go the next day. We woke up the next morning and found Digger with his front right leg sticking out at right angles to his body. He was going to stay with friends of ours while we were away, and Mum decided that we couldn't take him to them in the state he was in.

We took Digger to the vet, but as soon as we got into the surgery he decided that he didn't want to be there. Mum put him down and he ran for the car with all four legs on the ground!

He knew we wouldn't leave him if he had a poor sore paw!

Clare Rossiter
Connewarre, Victoria
Australia

A cat with a mind of his own

Our lilac point Siamese, Titus, was a cat who loved playing games, mostly of his own invention. A great favourite was to hide behind the end of the sofa, give the signal in a quavering howl, and leap out to ambush us as we obediently raced up the room.

At the end of our garden was an area of woodland and scrub where he loved to play a similar sort of hunting game. He would run off and hide, leap out to ambush us as we searched for him, then race away again to repeat the performance until we all ran out of steam. It must have looked extraordinary to passers-by to see the

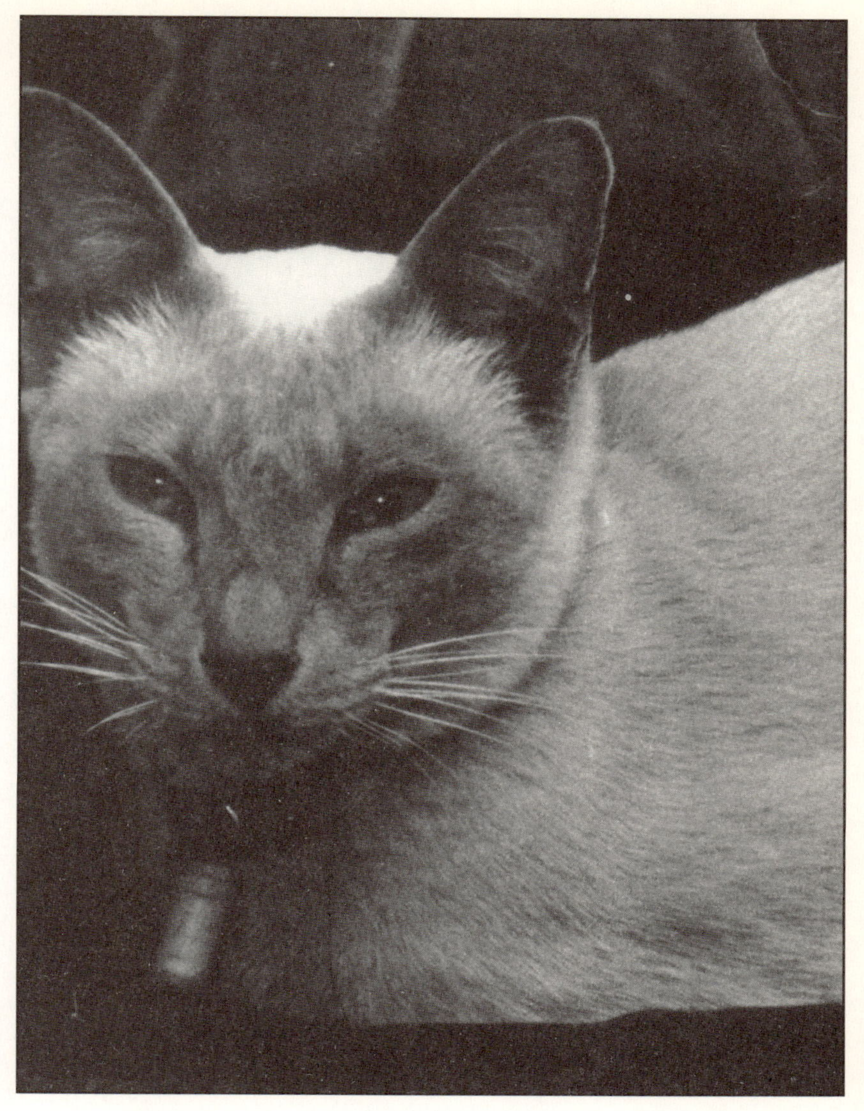

Titus loved playing games

two of us rushing up and down the woodland paths being hunted by a Siamese cat.

His other love was to play retrieving games with rubber thimbles, placed slightly out of reach in positions that would tax his ingenuity in fetching them. High spots were always a challenge to him.

Although generally well behaved, he was a decisive cat and did not take kindly to being opposed. As a recently arrived kitten, he jumped up onto the breakfast table and was promptly removed. He jumped up again, and a battle of wills ensued – which I lost. After the nineteenth time, I gave in, then built a wall of cereal cartons around my plate and finished my breakfast in a state of siege!

One night, we were woken by sounds of mayhem in the room downstairs. Convinced we were being burgled, we crept down to find a blackbird flying round the room, with Titus in hot pursuit across the furniture, knocking items flying as he went. He had very gently brought the blackbird in through his cat flap, and then released it so he could play with it in an enclosed space. The blackbird was quite unharmed and, when the front door was opened, flew away.

Siamese are, of course, renowned for their demanding conversation, and Titus was no exception. Whenever he came in through his cat flap he would wait for us to greet him. Any failure on our part was met by a hollow yowl of disapproval.

The cat flap itself gave us insight into his processes of deduction – or was it yet another chance to play a game? We had installed a sophisticated two-way design, advertised as being easy for the dimmest cat to master because the door was pushed forward in both directions. Not Titus. Having pushed forward on the way out, it struck him as only logical that the flap should be pulled towards him on the way in. It took him two days to master – or pretend to master – this piece of human trickery, which he clearly regarded with disfavour.

But one memory more than any other illustrates his impudence. The cling-clang-clonk of the cat flap on a cold wet night, followed by the soft padding of paws up the stairs, was as efficient a dream destroyer as any alarm clock. Titus was expert at finding his way beneath the duvet, and he had a set procedure for doing so. He would jump onto the pillow and pat a cheek with his cold paw. If we failed to respond, he would repeat the action with his claws out – then move on to a more determined assault on other parts of the anatomy.

Surrender was inevitable. Never pleased at being wet and cold, he would snuggle down against a warm body – heavy, soggy and determined. The electrifying shock of those encounters will never be forgotten.

Once in the bed, he would not be budged from human contact. At any attempt to move or roll away, he would simply follow and resettle. There were occasions when I woke up on the very edge of the bed – and once I moved away just a bit too far. Hitting the carpet with a heavy thump, I looked up to see a clearly astonished Titus staring down at me with a look that seemed to say, *And just what do you think you're doing down there?*

Colin Johnson
Brentwood, Essex
England

Write to me …
Email Colin
ctjohnson1943@aol.com

My way

Jaz, our Jack Russell terrier, is a master in the art of getting his own way.

He sees me as the 'pack leader', and when I am around he is usually obedient and reasonably subservient – for a Jack Russell,

that is! Jaz considers himself number two in the pecking order and my husband Brian a rather lowly number three. Using a variety of barks and growls and the occasional nose-bunt, Jaz manipulates Brian into doing his bidding on a daily basis.

He sleeps on top of our bed, usually on the outer edge beside Brian. One of his favourite early morning tricks, particularly if it is frosty outside, is to jump off the bed, take up his 'Do as I say' stance beside Brian and start to whine. If his message is ignored, Jaz's whining begins to sound more irritated and forceful. He has learned from experience that this approach has a 100 per cent success rate.

Sooner or later, Brian will open an eye and say, 'Do you need to go outside, boy?' To this, Jaz always replies with one enthusiastic woof, whereupon Brian climbs out of our lovely warm bed to let the dog out.

With his back turned to the bed, Brian barely has time to reach the door before Jaz is back up on the bed in one leap and curled up where he had wanted to be all the time – right beside me!

Cheryll Gadsby
Hawera, Taranaki
New Zealand

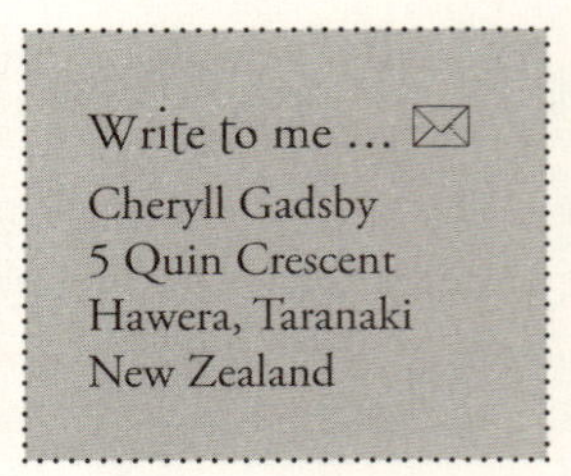
Write to me …
Cheryll Gadsby
5 Quin Crescent
Hawera, Taranaki
New Zealand

You can't stop me

Our two-year-old cat Pan hates lawnmowers. Every time she hears one at our place or at any of the houses around us, she runs out the cat door in the kitchen in a panic. I don't know where she goes when a lawnmower starts up, and I worry that she might not be safe.

Pan with a little koala friend

One time when I saw the lawnmowing people arrive, I decided to shut her inside. Since her cat door is permanently set to open, I shut the sliding door to the kitchen. It's one of those doors on rollers that sits closed and you push it along to open it.

When the lawnmower started up, Pan looked at me angrily for a minute, then went to the far end of the sliding door, jumped up on her hind legs with her front legs on the edge of the door, pushed it open, and bolted into the kitchen and out the cat door as usual.

Gina Sturkenboom
Hamilton
New Zealand

Write to me ...
Email Gina
gas4@waikato.ac.nz

Size doesn't matter

I leave wild birdseed out each day for the birds who 'pop in'. One day I was watching a tiny bird eating from the seed dish, all alone. Suddenly, several crimson rosellas and a couple of galahs flew down and took over the seed dish.

The little bird hopped down to the ground and pecked about for a while, seemingly unperturbed. Suddenly he flew straight up, scaring the larger birds into flight. He then resumed his solitary feast.

Joyce Elphick
Stanthorpe, Queensland
Australia

An escape artist

I acquired Duke, a two-year-old German Shepherd, from a couple whose relationship had ended; sadly, he was part of the 'marital debris'.

The day I brought him home with me, he was so excited he leaped over every wall in the garden. *Boing!* Over into my neighbour's garden. *Boing!* Over into the field. *Boing!* He scrambled up and over a big five-bar gate. Fortunately, he bounded back to me. He was like a canine kangaroo!

Apart from being very highly strung and in possession of a certain obstinacy, Duke came to me in a very sorry state, both physically and mentally. For one thing, he was painfully thin. I used to get embarrassed taking him for walks, in case people mistakenly believed me to be the cause of his skinny condition. Fortunately he soon put on weight.

I often wished Duke could tell me his story, but as time went on his habits gave clues to his previous existence. Duke loved company, especially women, and grew immensely excited around horses and children. He hated wasps with an intensity that bordered on psychotic! Most of all, though, he hated being locked in or restrained. Apparently, his former life had involved him being chained up for hours in a stable (with only wasps for company?) while the proper residents, the horses, were away grazing in a field.

He demonstrated his escape skills on many occasions – often to my embarrassment. Once, he freed himself from being tethered outside and found his way down into the cellar of the pub while my friend and I were ordering drinks. Despite all the steaks for the restaurant's evening meals being stored in the cellar's coolness, he never sniffed or touched a single one, so intent was he on seeking me out. We were reunited by a particularly red-faced and startled-looking chef, who was nervously shepherding my dog – from a very safe distance behind him!

The best of all his escape stories, though, was one I did not bear witness to but which came to me from a very reliable source, a boarding kennel owner who had served as a metropolitan police dog handler for many years. I had put Duke in his care while I visited my parents – and never in all his years on the force had he come across a character like Duke!

Duke seemed to value company as much as he valued a bowl of food. So intent was he on retaining companionship that he would try to grab the ex-police officer's arm and hold on to him to keep him there beside him. He never hurt him, though. When that failed to do the trick and the man tried to make his escape, Duke would hook a paw around his new friend's leg – almost tripping him up on more than one occasion.

So keen was Duke to have human companionship that he kept finding his way out of his kennel enclosure, managing to get though two latches to do so. Each time he escaped (apart from one night

when he scaled a seven foot outer perimeter fence, causing a lengthy search late at night) he would sit outside the back door of the house barking to be let in. Eventually they conceded defeat, and Duke was the only boarder granted permission to move in with his sitters!

The kennel owner was puzzled by Duke's ability to release himself from the enclosure, so he lay in wait one evening after feeding time, knowing that Duke would be sitting outside his back door before too long. Out of sight, he watched in amazement as Duke, with practised skill, slid the bolt across with his mouth, then lifted the latch with one paw and pulled the door open with the other! Quite a complicated series of actions. That was a few years ago, but I bet the kennel owner still remembers him clearly.

Duke is gone now, having died at 13 years old, but I'll never forget him. He was one of the best friends I've ever had.

Liz Hipkins
Walsall, West Midlands
England

Write to me …

Liz Hipkins
23 Holtshill Lane
Walsall, West Midlands
WS1 2JA
United Kingdom

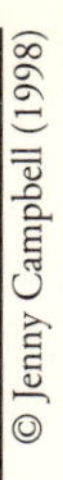

Here's a cheeky photo: Flash gives reading a try

Like what you're reading? Tell us your favourite story

We hope you've enjoyed reading about smart animals. We'd love to know which story is your favourite. This will help us choose the stories for future 'best of' editions of SMARTER than JACK.

Please write the book name, page number and story title on the back of an envelope or postcard. You will go into the draw to win a one-year subscription to the SMARTER than JACK series. There will be a draw every time a new book is released and the winner will be announced on our website.

For submission information please go to page 146.

The SMARTER than JACK story

We hope you've enjoyed this book. The SMARTER than JACK books are exciting and entertaining to create and so far we've raised over NZ$340,000 to help animals. We are thrilled!

Here's my story about how the SMARTER than JACK series came about.

Until late 1999 my life was a seemingly endless search for the elusive 'fulfilment'. I had this feeling that I was put on this earth to make a difference, but I had no idea how. Coupled with this, I had low self-confidence – not a good combination! This all left me feeling rather frustrated, lonely and unhappy with life. I'd always had a creative streak and loved animals. In my early years I spent many hours designing things such as horse saddles, covers and cat and dog beds. I even did a stint as a professional pet photographer.

Then I remembered something I was once told: do something for the right reasons and good things will come. So that's what I did. I set about starting Avocado Press and creating the first New Zealand edition in the SMARTER than JACK series. It was released in 2002 and all the profit went to the Royal New Zealand SPCA.

Good things did come. People were thrilled to be a part of the book and many were first-time writers. Readers were enthralled and many were delighted to receive the book as a gift from friends and family. The Royal New Zealand SPCA was over $43,000 better off and I received many encouraging letters and emails from readers and contributors. What could be better than that?

How could I stop there! It was as if I had created a living thing with the SMARTER than JACK series; it seemed to have a life all of its own. I now had the responsibility of evolving it. It had to continue to benefit animals and people by providing entertainment, warmth and something that people could feel part of. What an awesome responsibility and opportunity, albeit a bit of a scary one!

It is my vision to make SMARTER than JACK synonymous with smart animals, and a household name all over the world. The concept is already becoming well known as a unique and effective way for animal welfare charities to raise money, to encourage additional donors and to instil a greater respect for animals. The series is now in Australia, New Zealand, the United States of America, Canada and the United Kingdom.

Avocado Press, as you may have guessed, is a little different. We are about more than just creating books; we're about sharing information and experiences, and developing things in innovative ways. Ideas are most welcome too.

We feel it's possible to run a successful business that is both profitable and that contributes to animal welfare in a significant way. We want people to enjoy and talk about our books; that way, ideas are shared and the better it becomes for everyone.

Thank you for reading my story.

Jenny Campbell

Jenny Campbell
Creator of SMARTER than JACK

Submit a story for our books

We are always creating more exciting books in the SMARTER than JACK series. Your true stories are continually being sought.

You can have a look at our website www.smarterthanjack.com. Here you can read stories, find information on how to submit stories, and read entertaining and interesting animal news. You can also sign up to receive the Story of the Week by email. We'd love to hear your ideas, too, on how to make the next books even better.

Guidelines for stories

- The story must be true and about a smart animal or animals.
- The story should be about 100 to 1000 words in length. We may edit it and you will be sent a copy to approve prior to publication.
- The story must be written from your point of view, not the animal's.
- Photographs and illustrations are welcome if they enhance the story, and if used will most likely appear in black and white.
- Submissions can be sent by post to SMARTER than JACK (see addresses on the following page) or via the website at www.smarterthanjack.com.
- Include your name, postal and email address, and phone number, and indicate if you do not wish your name to be included with your story.
- Handwritten submissions are perfectly acceptable, but if you can type them, so much the better.
- Posted submissions will not be returned unless a stamped self-addressed envelope is provided.
- The writers of stories selected for publication will be notified prior to publication.
- Stories are welcome from everybody, and given the charitable nature of our projects there will be no prize money awarded, just recognition for successful submissions.

- Particpating animal welfare charities and Avocado Press have the right to publish extracts from the stories received without restriction of location or publication, provided the publication of those extracts helps promote the SMARTER than JACK series.

Where to send your submissions

Online

- Use the submission form at www.smarterthanjack.com or email it to submissions@smarterthanjack.com

By post

- **In Australia**
  PO Box 170, Ferntree Gully, VIC 3156, Australia
- **In Canada and the United States of America**
  PO Box 819, Tottenham, ON, L0G 1W0, Canada
- **In New Zealand and rest of world**
  PO Box 27003, Wellington, New Zealand

Don't forget to include your contact details. Note that we may use the information you provide to send you further information about the SMARTER than JACK series. If you do not wish for us to do this, please let us know.

Receive a SMARTER than JACK gift pack

Did you know that around half our customers buy the SMARTER than JACK books as gifts? We appreciate this and would like to thank and reward those who do so. If you buy eight books in the SMARTER than JACK series we will send you a free gift pack.

All you need to do is buy your eight books and either attach the receipt for each book or, if you ordered by mail, just write the date that you placed the order in one of the spaces on the next page. Then complete your details on the form, cut out the page and post it to us. We will then send you your SMARTER than JACK gift pack. Feel free to photocopy this form – that will save cutting a page out of the book.

Do you have a dog or a cat? You can choose from either a cat or dog gift pack. Just indicate your preference.

Note that the contents of the SMARTER than JACK gift pack will vary from country to country, but may include:

- The SMARTER than JACK mini Collector Series
- SMARTER than JACK postcards
- Soft animal toy
- Books in the SMARTER than JACK series

Show your purchases here:

Book 1	Book 2	Book 3	Book 4
Receipt attached ☐ *or* Date ordered ________	Receipt attached ☐ *or* Date ordered ________	Receipt attached ☐ *or* Date ordered ________	Receipt attached ☐ *or* Date ordered ________
Book 5	**Book 6**	**Book 7**	**Book 8**
Receipt attached ☐ *or* Date ordered ________	Receipt attached ☐ *or* Date ordered ________	Receipt attached ☐ *or* Date ordered ________	Receipt attached ☐ *or* Date ordered ________

Complete your details:

Your name: ______________________________

Street address: ______________________________

City/town: ______________________________

State: ______________________________

Postcode: ______________________________

Country: ______________________________

Phone: ______________________________

Email: ______________________________

Would you like a cat or dog gift pack? CAT/DOG

Post the completed page to us:

- **In Australia**
  PO Box 170, Ferntree Gully, VIC 3156, Australia
- **In Canada and the United States of America**
  PO Box 819, Tottenham, ON, L0G 1W0, Canada
- **In New Zealand and rest of world**
  PO Box 27003, Wellington, New Zealand

Please allow four weeks for delivery.

Which animal charities do we help?

At SMARTER than JACK we work with many charities around the world. Below is a list of some of the charities that benefit from the sale of our books. For a more complete list please visit www.smarterthanjack.com. If would like your charity to benefit from SMARTER than JACK please contact us by email: info@smarterthanjack.com.

New Zealand

Royal New Zealand SPCA and their branches and member societies: www.rspca.org.nz

Australia

RSPCA Australia and their eight state and territory member Societies: www.rspca.org.au

United Kingdom

Several animal welfare organisations in the United Kingdom

Canada

The Canadian Federation of Humane Societies and their participating member societies: www.cfhs.ca

United States of America

Around 25 humane societies, welfare leagues and SPCAs from all over the United States of America

Get more wonderful stories

Now you can receive a fantastic new-release SMARTER than JACK book every three months. That's a new book every March, June, September and December. The books are delivered to your door. It's easy!

Every time you get a book you will also receive a copy of our members-only newsletter. Postage is included in the subscription price if the delivery address is in Canada, the United Kingdom, Australia or New Zealand.

You can also purchase existing titles in the SMARTER than JACK series. To purchase a book go to your local bookstore or visit our website **www.smarterthanjack.com** and select the participating charity that you would like to benefit from your purchase.

How your purchase will help animals

The amount our partner animal welfare charities receive varies according to how the books are sold and the country in which they are sold. Contact your local participating animal welfare charity for more information.